SpringerBriefs in Materials

Series Editors

Sujata K. Bhatia, University of Delaware, Newark, DE, USA

Alain Diebold, Schenectady, NY, USA

Juejun Hu, Department of Materials Science and Engineering, Massachusetts Institute of Technology, Cambridge, MA, USA

Kannan M. Krishnan, University of Washington, Seattle, WA, USA

Dario Narducci, Department of Materials Science, University of Milano Bicocca, Milano, Italy

Suprakas Sinha Ray ⑩, Centre for Nano-Structured Materials, Council for Scientific and Industrial Research, Brummeria, Pretoria, South Africa

Gerhard Wilde, Altenberge, Nordrhein-Westfalen, Germany

The SpringerBriefs Series in Materials presents highly relevant, concise monographs on a wide range of topics covering fundamental advances and new applications in the field. Areas of interest include topical information on innovative, structural and functional materials and composites as well as fundamental principles, physical properties, materials theory and design. **Indexed in Scopus (2022).**

SpringerBriefs present succinct summaries of cutting-edge research and practical applications across a wide spectrum of fields. Featuring compact volumes of 50 to 125 pages, the series covers a range of content from professional to academic. Typical topics might include

- A timely report of state-of-the art analytical techniques
- A bridge between new research results, as published in journal articles, and a contextual literature review
- A snapshot of a hot or emerging topic
- An in-depth case study or clinical example
- A presentation of core concepts that students must understand in order to make independent contributions

Briefs are characterized by fast, global electronic dissemination, standard publishing contracts, standardized manuscript preparation and formatting guidelines, and expedited production schedules.

Serhii Sheyko · Yurii Belokon · Oleksii Hrechanyi ·
Tetyana Vasilchenko

Thermoplastic Processing of Structural Metallic Materials

Experiments, Theory, and Modeling

Springer

Serhii Sheyko ⓘ
Head
of the Educational-Scientific-Production
Center "Metalspetsproekt"
Zaporizhzhya National University
Zaporizhzhia, Zaporiz'ka, Ukraine

Oleksii Hrechanyi ⓘ
Department of Metallurgical Equipment
Zaporizhzhia National University
Zaporizhzhia, Zaporiz'ka, Ukraine

Yurii Belokon ⓘ
Department of Metallurgical Technologies
Ecology and Technogenic Safety
Zaporizhzhia National University
Zaporizhzhia, Zaporiz'ka, Ukraine

Tetyana Vasilchenko ⓘ
Department of Metallurgical Equipment
Zaporizhzhia National University
Zaporizhzhia, Zaporiz'ka, Ukraine

ISSN 2192-1091   ISSN 2192-1105 (electronic)
SpringerBriefs in Materials
ISBN 978-3-031-73895-1   ISBN 978-3-031-73896-8 (eBook)
https://doi.org/10.1007/978-3-031-73896-8

This Springer imprint is published by the registered company Springer Nature Switzerland AG
The registered company address is: Gewerbestrasse 11, 6330 Cham, Switzerland

If disposing of this product, please recycle the paper.

Preface

The development of metal processing by pressure method, the introduction and mastering of new high-performance equipment, rolling mills, entire complexes of processing units, continuous expansion of the assortment of alloys that are subject to hot plastic processing, require continuous improvement and refinement of technological process design methods. To solve these issues, it is necessary to study the mechanical properties of alloys under different temperature–speed deformation conditions and develop reliable methods of their assessment in specific technological processes to ensure maximum equipment productivity and improve product quality.

During the 2000s, most of the researchers of GKSS (Germany) and the Frantsevych Institute for Problems of Materials Science (Ukraine) recognized that in order to ensure the desired combination of properties of special alloys—high heat resistance, plasticity, and fracture toughness, it is necessary to have an equiaxed lamellar structure with a small grain size. Reducing the grain size by thermomechanical treatment to the level of d $\approx$ 10 μm leads to a further increase in plasticity at room temperature, but at the same time there is a shift in the temperature of the brittle-viscous transition from $\approx$ 800 to 600...700 °C and a corresponding decrease in the temperature of potential using such alloys. The problem of grinding the structure of ingots of refractory alloys is crucial, since only by achieving a homogeneous structure with a relatively small grain size in the ingot can one expect to improve the technological properties. As noted by leading scientists in the field of obtaining intermetallic alloys, Imaiev V. M. and G. Clemens, thermoplastic processing technology can be very promising from a practical point of view, since such alloys belong to hard-to-deform and low-tech materials. In particular, the promising application of powder technologies is presented in relation to the production of semi-finished products from γ-TiAl alloys.

However, due to the peculiarities of powder technologies, the blanks obtained by this technology are often characterized by the segregation of chemical elements and the presence of pores in the volume, which requires further technological operations aimed at eliminating these shortcomings. A promising method of eliminating these shortcomings is the use of thermochemical pressing, which can be applied to a wide class of materials and allows to simultaneously grind the material structure and increase the homogeneity of the alloy. Grinding of the intermetallic alloy grain in the process of its synthesis under pressure occurs as a result of plastic deformation of the synthesis product and high cooling rates. A higher efficiency of the process of grinding the grain of the intermetallic alloy structure can be achieved with thermoplastic deformation of the synthesized alloy in the process of forming the grain of the structure during the high-temperature synthesis of the alloy under pressure. Therefore, an increase in the temperature of deformation under the conditions of thermochemical pressing and a decrease in the grain size and the share of twin grain boundaries lead to the delocalization of deformation due to a change in the nature of sliding, as well as features of the interaction of dislocations with grain boundaries and an increase in plasticity.

The main purpose of writing our monograph is to research ways of obtaining special-purpose alloys with improved technological properties by methods of thermoplastic deformation, which will be interesting not only to technologists of manufacturing enterprises, but also to students and researchers in the field of materials science and metallurgical equipment.

Zaporizhzhia, Ukraine

Serhii Sheyko
Yurii Belokon
Oleksii Hrechanyi
Tetyana Vasilchenko

Acknowledgements

We had the idea of writing this book even before the war in our country. Constant sounds of air raids and explosions outside the window, frequent power outages did not stop us, but only hardened our character and accelerated work on the book. Thanks to the strong support of our Western partners, Ukraine continues to surprise the whole world. We hope that our scientific work will be useful not only to experience material science practitioners, but also to enthusiastic researchers from all over the world who are just taking the first steps in great science.

We would like to express special thanks to our manuscript editors, Zachary Evenson and Ravi Vengadachalam, who supported us during the writing of the manuscript and provided valuable suggestions to improve it. To appreciate the contribution of these professionals, it is worth noting that if you are reading this book now, it is 50% their merit.

And finally, I would like to add that no matter how serious the material in scientific literature is, there will always be a place for a smile in life. We wish you all the best, dear reader.

Zaporizhzhia, Ukraine
August 2024

Contents

Chapter 1
Current State of Research and Prospects for the Development of Thermoplastic Processing of Dual-Phase Alloys

The development of technology for the treatment of metals by pressure, the introduction and development of the new high-performance equipment, rolling mills, whole complexes of processing units, continuous expansion of the assortment of steels and alloys that undergo hot plastic processing, require continuous improvement and refinement of methods of designing technological processes. To solve these problems, it is necessary to study the mechanical properties of steels and alloys at various temperature-speed deformation conditions and develop reliable methods for their evaluation in specific technological processes to ensure maximum productivity of equipment and product quality improvement.

The accuracy of the calculations depends on the reliability of the determination the current flow of metal. The stress of the metal flow is the main component, which has a direct proportional effect on the accuracy of the calculation of power-supply parameters of rolling. In studies, it was found that the stress of the metal flow during hot plastic deformation depends on the temperature, degree and rate of deformation. The temperature of the metal is the most powerful factor that determines the magnitude of the current stress σy. As the heating temperature rises, all the characteristics of the metal are reduced. An increase in the degree of deformation leads to an increase in the stress flow, and as a result - the metal is further strengthened. The stress of the flow increases significantly with the increase in the rate of deformation of the metal.

Therefore, the development of models for determining the metal flow in hot rolling processes is an urgent task in the theory of rolling and requires a detailed study.

S. Sheyko et al., *Thermoplastic Processing of Structural Metallic Materials*, SpringerBriefs in Materials, https://doi.org/10.1007/978-3-031-73896-8_1

1.1 State of the Problem of Optimizing the Hot Deformation Modes of Dual-Phase Alloys

The tension of the metal flow is an important physical quantity characterizing the plastic deformation of metals and alloys. The stresses of the metal flow are used in the calculations of the power-supply parameters of the processes of metal-pressure treatment, which must be carried out in the case of developing new and improving existing technological processes, using new grades of steels and alloys, with the choice and calculation of the equipment, etc.

To date, sufficiently large experimental material has been accumulated on the stresses of steels at different temperatures, degrees, and rates of deformation using various installations and test methods, including modern testing machines such as cam-plastmeters.

The experimental studies of the following authors are known: A. Nadai, L. Sokolova, A. Dinnika, A. Chekmareva, I. Tarnovsky, S. Gubkina, V. Zyuzin, M. Brovman, L. Andreyuk and G. Tyuleneva, P. Polukhin G. Guna, P. Cook, Suzuki and others.

Many curves of the metal flow under deformation are given in the scientific and reference literature [1, 2].

We note the work of S. Gubkin, dedicated to the plastic deformation of metals. The author proposed a series of mathematical models of rheology for describing the curves of the metal flow under deformation relatively to different directions of rheology: the theory of flow, the theory of aging, the theory of deterioration, and others.

V. Whitman and M. Zlatin proposed a formula that takes into account the influence of epy temperature and velocity of deformation. A. Tselikov and V. Persiantsev offered an equation for determining the stress of the metal flow as a result of the simultaneous action of two processes: strengthening and deterioration at a constant rate of deformation. Later, this method was developed by V. Persiantsev for the case of metal deformation with variable speed. L. Sokolov proposed an equation describing a change in the deformation resistance, which takes into account additional relaxation of the stress in addition to its release. Y. Schwartzbarth derived an equation for determining the current stress in an isothermal process using a sequential superposition of simple functions of strengthening and debilitating (relaxation). According to Y. Schwartzbart "the validity of the accepted principle of superposition is almost entirely satisfactory with the convergence of the received function by known curves of the current." He introduced the notion of the characteristic magnitude of the deformation $\varepsilon = u/A$, corresponding to the maximum of the deformation curve.

In the monograph, I. Tarnovsky, V. Eremieva, C. Baakashvili, outlined the methods of calculating the resistance of metal deformation using the theory of creep. M. Zaykov gives formulas for calculating the current stress from the temperature, degree and deformation rate obtained with the application of the basic provisions of the theory of stressed state and the thermodynamics of irreversible processes.

Using the thermomechanical parameters, the calculated formulas for the metal flow are obtained in the following works: V. Zyuzin, M. Browman, L. Andreyuk and G. Tyuleneva, V. Nikolaev.

Modern works are aimed at obtaining experimental curves of the metal flow under high plastic deformations by the torsion plastometry method. Different analytical dependencies are proposed that describe all types of curves of high-temperature deformation hardening. An overview and analysis of mathematical models of rheology is given in [1-8, etc.].

A large volume of background information on the stresses of the flow of various metals and alloys is given in the works of P. Klymenko. In his works P. Klimenko during the analysis of plastic deformation resistance introduces the concept of "relative strengthening". The dimensionless curve of relative strength change can be described by a single equation. Using this equation, it is easy to proceed to the construction of the plastometric curves of the current stress in the form of a change in the absolute values of the stress.

For the case of hot rolling, the term "relative stress flow" is introduced. The analytical equations that describe the change in the relative stress of the flow along the deformation cell are presented. Using this equation, it is possible to obtain a distribution in the deformation cell of the current stress.

Thus, the stress of the metal flow is the main component, which has a direct proportional effect on the accuracy of the calculation of power parameters of rolling. As can be seen from the review of the state of the problem, the determination of the metal flow is a complicated problem in the metal abandonment theory. Therefore, the development of models for determining the metal flow in hot rolling processes is an urgent task in the rolling theory and requires a detailed study.

The stresses of the metal flow are the stresses of the flow of metal under static conditions of plastic deformation [1].

The fluidity stress characterizes the mechanical properties of the material in the headquarters in the sections immediately before and after the deformation, as well as in the intervals between the cages and in pauses between passes during rolling, respectively, in continuous and reversible states. Therefore, the establishment of regularity changes in the stress of the flow of metal headquarters is important not only for the theory of rolling, but also for technology.

The theoretical determination of the current of metal current is impossible at present. Therefore, data on the magnitude of this parameter are obtained only experimentally. In most cases, the stresses of the metal flow are judged by the results of testing the samples for stretching.

The stresses of the metal flow depend on the chemical composition and structure parameters, that is, on the physical and chemical properties of the deformed metal (alloy), the previous and the partial relative compression, the temperature and the fineness of the deformation, as well as from the stress state circuit, technological lubrication and some other less significant factors. However, the chemical composition (physical and chemical properties) of the deformed metal, the total relative bending and the deformation temperature have the main influence on the stresses of the metal flow [1].

Among the chemical elements that are part of carbon steels, carbon has the greatest influence on the stresses of the metal flow. With the increase of carbon content in the steel, the stress of the flow increases. The alloying elements are also contribute to the increase of the stresses of the metal flow of the steel. A similar relationship between the stresses of the metal flow and the alloying elements is also a characteristic of other metal alloys.

For a general type of curve, a stream of steels shown in Fig. 1.1 is the characteristic growth σ_y from the yield curve σ_0 to a certain peak value σ_p corresponding to the peak deformation ε_p, after which σ_τ smoothly decreases to the value corresponding to the constant stress σ_y, at which the equilibrium of the processes of strengthening and dynamic recrystallization comes in. It should be noted that at the site corresponding to the stresses from σ_0 to σ_p, the hardening rate of the metal decreases as a result of the prevailing process of dynamic return over the strengthening. Dynamic recrystallization begins when the deformation ε_x reaches. The nature of the curve of the stream also reflects the change of austenitic grain. In the area of hardening, until the deformation reaches the value ε_p, the grain is crushed, the density of the dislocation of the substructure increases, after which the process of dynamic recrystallization intensively develops, and according to the work [3], the grain size of austenite at the established voltage depends exclusively on σ_y.

Proceeding from the preconditions made in [5, 9], using the principle of superposition of strengthening and defeat processes, in [11] an expression was obtained for the instantaneous deformation resistance:

$$\sigma_s = \sigma_y + \left(\sigma_0 - \sigma_y + D \cdot \varepsilon\right) \exp(-\varepsilon/\varepsilon_x), \tag{1.1}$$

where D is the parameter characterizing the rate of hardening of the metal.

The parameters σ_0 and D are difficult to determine when processing experimental data, since the flow curve at the initial moment strongly depends on the structure of the metal. On the other hand, existing plastmets do not allow to fix the initial stage of the flow of metal. The analysis of the experimental curves of the flow shows that the maximum resistance on the curve is the maximum resistance σ_p and the logarithmic

Fig. 1.1 General view of the curve of a steel stream in the presence of dynamic recrystallization

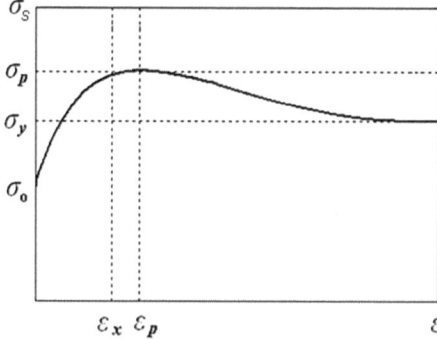

deformation ε_p of the maximum strain hardening, respectively. Therefore, the expression (1.1) has been transformed by eliminating difficult identifying parameters from it. Finally, the following expression was obtained for σ_s:

$$\sigma_s = \sigma_y + (\sigma_p - \sigma_y) \cdot \left(\frac{\varepsilon - \varepsilon_p}{\varepsilon_x} + 1 \right) \exp\left(\frac{\varepsilon_p - \varepsilon}{\varepsilon_x} \right), \tag{1.2}$$

To calculate the stresses of the metal flow according to the proposed model it is necessary to determine the thermokinetic dependences of the parameters σ_y, σ_p, ε_p and ε_x, taking into account the chemical composition of the steel.

As a result of plastic deformation there is a strengthening of the metal, therefore, with the increase of the partial and total relative compression during the rolling of the stress of the flow of metal increases. Hot rolling is carried out at temperatures exceeding the temperature of recrystallization. For this reason, the metal in the process of hot rolling not only strengthens, but also becomes shortened. However, in general, the hot rolling process is characterized by strengthening the metal at the outlet of the deformation cell.

According to the theory of A. Nadai [10], the growth of the current stress for each metal (alloy) is expressed by the equation:

$$d\sigma = \frac{\partial \sigma_y}{\partial T} dT + \frac{\partial \sigma_y}{\partial \varepsilon} d\varepsilon + \frac{\partial \sigma_y}{\partial u} du + \frac{\partial \sigma_y}{\partial \tau} d\tau, \tag{1.3}$$

where

σ_y current of the metal;
T temperature;
ε relative deformation;
τ time of deformation;
u the rate of deformation.

The first three annexes of this equation take into account the influence on the stresses of the metal flow in relation to temperature, relative degree and rate of deformation, the latter - the law of the development of deformation in time.

The regularities necessary for the solution of the A. Nadai equation have not yet been sufficiently studied, therefore, practically the values of the stress of a metal flow are determined using the results of the corresponding experiments.

Experimental studies by definition of the stress of the flow of metal can be divided into two groups:

(a) Studies based on the measurement of the force acting on the pressure screws of the state when rolling headquarters with different values of degree (ε), temperature (T) and rate (u) of deformation;
(b) Studies based on the measurement of loads on test vehicles of a special design when stretching or compressing samples with different values of the parameters ε, T and u. The appearance of cam and torsion plastometres has expanded the

possibilities for an experimental study of the flow of metal and greatly facilitated the task of modeling the influence of laws of deformation development in time, in relation to specific processes of plastic deformation.

Parameters ε, T and u have different quantitative influence on the stresses of the metal flow during hot and cold rolling. The quantitative influence of the deformation rate on σt under hot strain is much greater than that of the cold. This is evidenced by the experimental dependences $\sigma_y = f(u)$ (Fig. 1.2) Dinnik [4] obtained by precipitation of samples from steel 3 (Standard of Ukraine) with a temperature of 600–1200 °C. The graphs $\sigma_y = f(u)$ given in this figure are constructed with $\varepsilon = 30\%$. To take into account the effect of actual compression on σt use a correction factor whose value is determined by means of an auxiliary graph:

$$\sigma_y = k \cdot \sigma_{y30},$$

where σ_{y30} is the stress of the metal current at compression, which is equal to 30%.

At active hot rolling mills, the deformation rate is in the range of 0.5–10 s^{-1} to 500–10^3 s^{-1}. As it follows from Fig. 1.2, an increase in the rate of deformation to the indicated values leads to an increase in the stress of the flow of metal in 2–4 times.

Let us also dwell on the experimental data of Cook [11] presented in the form of charts in the coordinates $\sigma_y - \ln \frac{h_0}{h_1}$ (Fig. 1.3). These data were obtained on a plastometer in conditions that ensure the constancy of the deformation rate with the

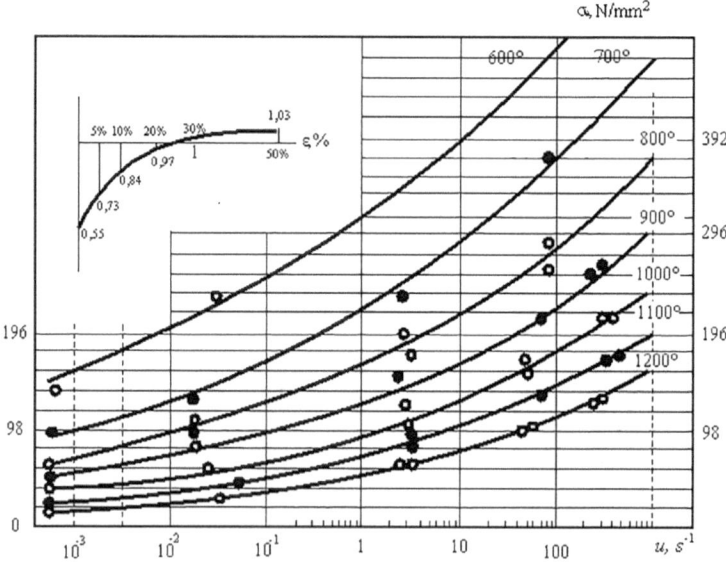

Fig. 1.2 Experimental dependences $\sigma_y = f(u)$ for hot deformation of steel 3 (Standard of Ukraine) (data of A. Dinnik)

Fig. 1.3 Experimental
dependencies $\sigma_y = f\left(\ln\frac{h_0}{h_1}\right)$
for hot deformation of low
carbon steel (0.15% C,
0.68% Mn) (data of P. Cook)

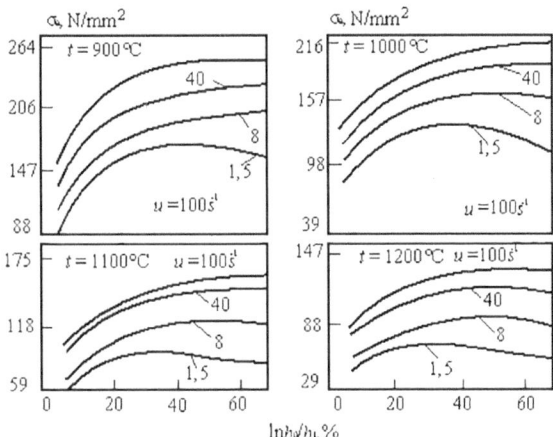

decrease in height and the preservation of the cylindrical shape of the deposited sample, that is, in conditions close to the linear stressed state.

The graphs $\sigma_y = f\left(\ln\frac{h_0}{h_1}\right)$ (Fig. 1.3) reflect the quantitative effect of strengthening the metal on σ_y at $t = const$ and $u = const$. The presence of a maximum on these curves shows that with a further increase in the actual relative bending, the prevailing influence on σ_m gives the resistivity of the metal.

Of the presented in Fig. 1.3 graphs shows that the decrease in the temperature of the deformation of carbon steels from 1200 to 900 °C for $u = const$ and $\varepsilon = const$ leads to an increase in the current flow of metal in 1.5–2.0 times, and an increase in relative compression from 5–10% to 30–40% for $t = const$ and $u = const$ causes an increase of σ_y by 40–60%.

A large complex of studies on the establishment of regularities of change σt for hot deformation is made by Zyuzinym [13]. They obtained quantitative data about σt for hot strain of 44 grades of steels and 20 non-ferrous metals and alloys. The results of these studies are published in the literature and will be discussed in more detail in the next section.

In the process of optimization from the point of view of product quality, relations that describe the conditions for material destruction are essential. From this point of view it is expedient to consider elements of a mathematical model that can be included in the system of constraints of the problem of optimization of the mode of changeover.

1.2 Characteristics of Methods for Determining the Stress of Metal Flow During Hot Deformation

There are several basic methods for determining the stress of the flow of metal experimentally—stretching, compression, torsion, method of reference pressure [3].

Stretching The stretch pattern is most often made in the form of a body of rotation with a cylindrical working part. During testing, in the working part of the sample, the stresses are constant and even irrespective of the mechanical properties of the material from which it is made. In tests, the average values of stress and deformation are measured - macroscopic quantities. At the same time, the unevenness of the deformation is approximately three times smaller than when compressed. In stretch marks, it is not possible to obtain a degree of deformation of more than 0.2–0.3. At large degrees of deformation, a neck is formed, after which the stress state is already heterogeneous and is not uniaxial stretching. The neck has a volumetric stressed state. The voltage in the neck increases with further growth of deformation, and in the rest of the total volume of the sample - decreases: there is an unloading. The sample stretches to fracture and determines the stresses corresponding to the maximum value of the force, that is, they find the boundary of strength, with curves similar to those shown in Fig. 1.4.

Compression The compression tests have the same principle as the stretch test, while this method has its drawbacks and advantages. A sample test is also made in the form of cylinders. Compression is carried out on presses, and there is friction between the ends of the sample and the working surface of the tool. Friction results in the so-called barrel-shaped specimen and volumetric stressed state, but it is very difficult to exclude the effect of friction during compression.

The values obtained by the experiment, as a rule, are overestimated. The advantage of the compression test process is the possibility of applying large deformation degrees. With large deformation values, the effect of friction on the deformation process decreases. To reduce the impact of friction, apply different lubricants. When

Fig. 1.4 Curves obtained during stretching tests

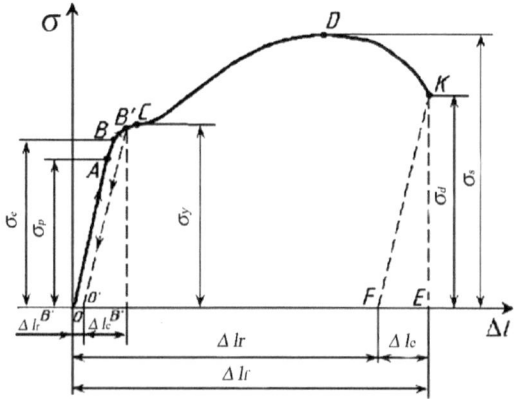

Fig. 1.5 Curves obtained during compression tests

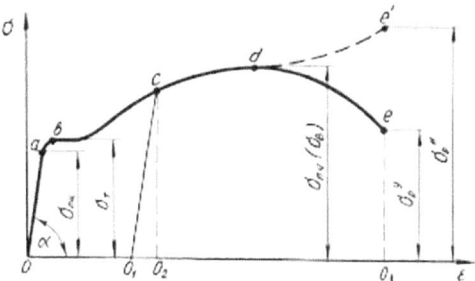

processing experimental data on the stress flow of the metal obtained in the case of a compression test, it is necessary to introduce a correction factor determined by comparing the results of testing the samples of the same size for stretching and compression; this coefficient takes into account the effect of friction on the contact surfaces. The type of curve obtained in compression tests is shown in Fig. 1.5.

Circumcision In addition to compression and stretching tests, the use of experiments on torsion of cylindrical specimens is also found [11]. The samples have a cylindrical working part and two thickened ends, which serve to fasten the sample in the car's grips. To the ends of the sample are applied two identical in magnitude torque, which cause the swirling of the sample at some angle. It is these two values that need to be measured. In the cylindrical part of the sample there is a tense state of pure shifts. However, the tense state is not homogeneous. The strain in the sample varies in radius according to the linear law. In the center of the sample, the voltage is zero. The largest value is observed on the surface of the sample. The following advantages of the torsion test are noted in the paper [9]: the constant of the deformation velocity for each individual load act, the absence of negative influence of contact forces of friction, the reach of high deformation levels, close to the exhaustion of the resource of the plasticity of the material under test, relative simplicity of the test installation.

The type of curve obtained during torsion tests is shown in Fig. 1.6.

Base specific pressure. Basic self-pressure method was proposed by O.I. Tselikovym. The essence of this method is that for the characteristics of the deformation resistance, the average specific pressure p0 at the rolling of reference samples

Fig. 1.6 Curves obtained during torsion tests

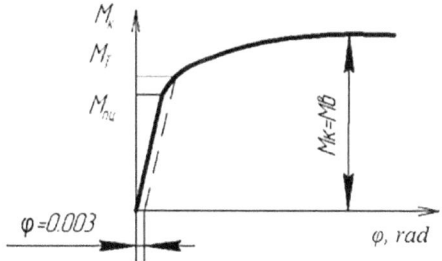

in the reference conditions is taken. According to the standard conditions, two-dimensional deformation conditions were adopted that eliminated the effect of expansion and minimized the effect of external friction and external zones. These conditions are largely adhered to when rolling samples of a rectangular cross section, when $b \gg H$ and $l = H$ so you can take $0 = 2 k$. The average specific pressure, determined at $b \gg H$, $l = H$, and the small value of the coefficient of friction, called the basic specific pressure, characterizes the resistance of the deformation of the metal in a flat-deformed state. The base unit pressure is recommended to be used instead of the value of 1.15 when calculating the pressure of the metal on the rolls by theoretical formulas.

Test on a plastometer. For data on the stresses of the metal flow, the most informative method is the sedimentation of samples on the plastometers [12], on which large degrees of deformation can be provided in a wide range of changes in temperature and deformation rates. In order to increase the accuracy of the reproduction of the given test modes, expansion of the range of modes and increase the reliability of test results on the basis of plastometers, an automated plastometric complex is created, which includes modern systems for recording and processing received data (for example, the installation of Gleeble). The use of specially designed computer programs for the processing of experimental data and the construction of curves of metal significantly increases the reliability and reliability of the results of experiments, provides the possibility of their storage in electronic form and the accumulation of data bank on the mechanical properties of the metals and alloys to be tested. Increased accuracy of the tests is also achieved thanks to the support of the speed at the lower level with an error of no more than 3% at any load. These measures allow obtaining curves of steels and alloys in a wide range of changes in degrees ε (5–80%), velocities u (0.01–100 s^{-1}) and temperatures (20–1300 °C) deformation at sample precipitation at the efforts of P in the process of deformation up to 500 kN. For tests on draft, cylindrical specimens are used that are related to the initial height ho to the diameter d_o in the range of 1.15–1.35. The most commonly used diameter of the samples is 12–15 mm. The height of the samples is determined by the geometry (profile) of the current cam.

The samples, as a rule, have end caps of height 0.5–0.7 mm for maintenance of lubrication. Before planting the sample for heating, they are rotated in pre-soaked sheet asbestos and placed in a special cylindrical container on the center of the siege plates—buoys from special heat-resistant alloys (Fig. 1.7).

Having analyzed the above methods of determining the magnitude of the stress of the flow of metal, we can draw the following conclusions: the methods of stretching and compression allow to determine the stresses of the flow of material under non-isterminous conditions, while the accuracy of determining the parameters of the graphs deformation-stress is small; the torsion method is more accurate in connection with the homogeneity of the stressed state, but the production of samples for this type of test is complicated; the method of base specific pressure is optimal for the case of application of the results of experiments in calculating the parameters of rolling, since in its reference conditions, the number of factors affecting the flow of metal, is minimized, and, consequently, the overall error in determining the average value of

Fig. 1.7 Container for sedimentation of samples on a plastometer [3]: 1—upper buoy, 2—lubricant, 3—test sample, 4—asbestos isolation, 5—container, 6—bottom boyok

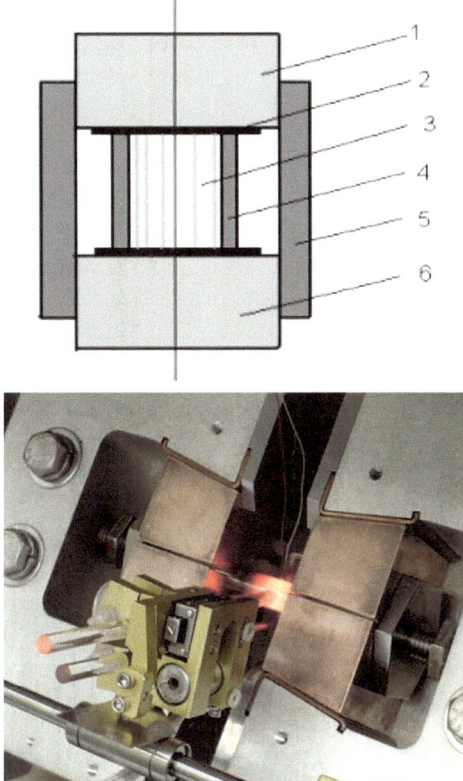

the stress flow at rolling out The disadvantage of the method is the impossibility of obtaining actual voltage values, which makes it difficult to apply the obtained data in the method of finite elements; tests on a plastometer allow data in a wide range of influencing parameters with a high degree of accuracy, but the conversion of values for different stress states reduces the accuracy of the result.

Thus, from the standpoint of the breadth of the research, the rational method is a test on a plastometer. In addition, when testing on these installations (Gleeble et al.), It is possible to accurately adjust and maintain the specified parameters (velocity, degree, temperature of deformation) during the test, therefore this method has become the most widely used in the study of different properties of metals.

1.3 Mechanisms of Hardening of Intermetallic Alloys and Dual-Phase Steels Under Hot Deformation

In the processes of modifying the hot deformation of dual-phase alloys, the following types of strengthening take place:

1. Hard-hardening strengthening due to alloying elements present in the alloy composition. The strengthening can be summed according to the equation:

$$\sum \sigma_{hh.s} = \sum_{i=1}^{n} K_i \cdot C_i$$

   where K_i is the ferrite reinforcement factor; C_i is the concentration of the element in the ferrite.

2. Grain-boundary hardening as a result of crushing of primary grain (dendrites) of δ-ferrite and austenite grain due to the formation of new crystallization centers based on dispersed nitrides and titanium carbonitrides. Strengthening is determined by the grain size of ferrite:

$$\sigma_g = K_y \cdot d^{-1/2}$$

   where K_y—coefficient of strengthening; d—grain size.

3. Strengthening due to the increase in the amount of perlite, associated with an increase in the stability of supercooled austenite due to modifying additives (titanium and aluminum):

$$\Delta \sigma_n = K_n \cdot \Delta P$$

   where K_i is the empirical coefficient is 2.4.

   ΔP—increase in the pearlite component during the modification, %.

4. Dispersion strengthening ($\Delta \sigma_{d.s.}$) of structural components due to the refractory and dispersed nitrides and carbonitrides of titanium and aluminum formed during the treatment with a modifier. This type of hardening is associated not only with the strengthening effect of the particles themselves, but also their interaction with dislocations. According to this model, dislocations in motion will be held on the particles until the applied stress is sufficient to cause the line of dislocations to bend and pass between the particles leaving a dislocation loop near them.

   Strengthening due to dispersed particles in dual-phase steels rises according to formula:

$$\Delta \sigma_{d.s.} = \frac{F \cdot G \cdot b}{2\pi(\lambda - 2r)},$$

where

r   particle radius;
λ   distance between particle centers;
F   coefficient characterizing the type of dislocations interacting with the particles
    (F = 1.25);
G   shear modulus;
b   Burgers vector.

5. Deformation strengthening ($\Delta\sigma_d$) as a result of an increase in the dislocation
    density, which is determined by the dependence:

$$\Delta\sigma_d = \alpha \cdot m \cdot G \cdot b \cdot \rho^{1/2},$$

where

α   coefficient of the character of the interaction of dislocations under strain
    strengthening;
τ   orientation multiplier (τ = 2.75); $\alpha \cdot m = 0.5$;
G   shear modulus (GFe = 84,000 MPa);
b   Burgers vector iron (b = 0.25 nm);
p   dislocation density.

Thus, the joint action of several hardening mechanisms can determine the total
yield strength, using the principle of E. Orovana according to the following formula

$$\sigma_y = \Delta\sigma_{hh.s.} + \Delta\sigma_g + \Delta\sigma_n + \Delta\sigma_{d.s.} + \Delta\sigma_d$$

Proceeding from the obtained data and estimating the contribution of each mechanism to the yield strength, it is possible to determine the total value of the yield
strength (Table 1.1).

The main types of strengthening of ferrite-pearlite steels are grain-boundary,
pearlite and deformation. Dispersion hardening causes strengthening not only due

Table 1.1 Characteristics of the yield strength of dual-phase steels after modification and deformation

Types of strengthening	Calculation formula for strengthening	Strengthening rate at yield strength (%)	Components that increase strengthening
Hard-hardening strengthening	$\sum \sigma_{hh.s} = \sum\limits_{i=1}^{n} K_i \cdot C_i$	25–40	Mn, Si (in solution)
Grain-boundary	$\sigma_g = K_y \cdot d^{-1/2}$	25–30	Ti, Al, N (in nitrides)
Perlite	$\Delta\sigma_n = K_n \cdot \Delta P$	10–15	Ti, Al, Ca (in solution)
Dispersion	$\Delta\sigma_{d.s.} = \frac{F \cdot G \cdot b}{2\pi(\lambda - 2r)}$	5–10	Ti, Al, N (in nitrides)
Deformation	$\Delta\sigma_d = \alpha \cdot m \cdot G \cdot b \cdot \rho^{1/2}$	15–25	Compression

to intrinsic contribution, but also indirect influence on grain hardening by grinding grain.

Thus, from the mentioned strengthening mechanisms acting on two-phase steels with different microstructures, it is possible to single out the dominant mechanism, which predetermines the efficiency of operation of metal structures. In steels with ferrite-pearlite structure it is a mechanism of grain-boundary, pearlite and strain hardening, which manifests itself to a considerable extent in steels after modifying treatment.

1.4   Models of Theoretical Determination of the Stress of Metal Flow During Hot Deformation

Using experimental data about σy presented in the form of graphs or tables, to calculate the current stress during hot rolling is inconvenient and leads to a decrease in the accuracy of the determination of the desired value. For ease of use of available experimental data about σ_y V. Zyuzin proposed a method of thermomechanical coefficients [13]. The essence of this method is that the rolling stress in a hot rolling is represented by both a product $\sigma 0$ and three independent coefficients. Mathematically σ_y by this method is written in the form:

$$\sigma_y = \sigma_0 k_t k_\varepsilon k_u, \tag{1.4}$$

where σ_0 is the basic value of the stress of yield; k_t, k_ε, k_u—coefficients taking into account the influence of temperature, degree and rate of deformation.

As the basic σ_0, the value of the stress of the flow of the material of the head is taken at $\varepsilon = 0.1$; $u = 10 \text{ s}^{-1}$; $T = 1000°C$.

The values of the thermomechanical coefficients k_t, k_ε, k_u for each metal (alloy) are determined according to the graphs. In Fig. 1.8 are graphs $k_t = f(T)$, $k_\varepsilon = f(\varepsilon)$, $k_u = f(u)$ for Steel 45 (Standard of Ukraine). The basic value of the stress current of this steel is 86 N/mm^2.

In [12] the basic values of the stress of σ_0 of some steels are given, as well as V. Zyuzine ranges of parameters ε, T, and u, which determine the scope of the model (2.1).

The application of the model (1.4) simplifies and accelerates the procedure for determining the current stress according to available experimental data. However, it does not rule out the need to use graphical dependencies to find the values of the coefficients k_t, k_ε, k_u.

The analysis of experimental dependences $k_t = f(t)$, $k_\varepsilon = f(\varepsilon)$, $k_u = f(u)$, V. Zyuzin, performed by Klimenko [14], showed that they are accurately approximated by the equations with acceptable accuracy for engineering practice:

at $\varepsilon < 0.15$

Fig. 1.8 Graphic
dependences for the
determination of the
thermomechanical
coefficients of Steel 45
(Standard of Ukraine) (data
of V. Zyuzin's)

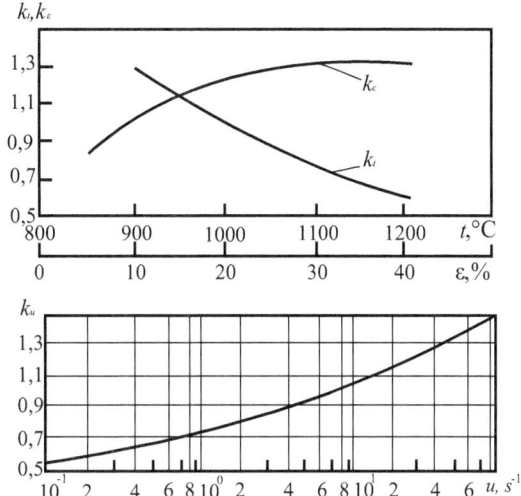

Fig. 1.8 Graphic
dependences for the
determination of the
thermomechanical
coefficients of Steel 45
(Standard of Ukraine) (data
of V. Zyuzin's)

$$k_\varepsilon = 4,7\sqrt{\varepsilon} - 4,5\varepsilon; \tag{1.5}$$

at $\varepsilon > 0.15$

$$k_\varepsilon = 0,82\left(1 + \sqrt{\varepsilon}\right); \tag{1.6}$$

at $u = 1 - 100 \text{ s}^{-1}$

$$k_u = 0,8 + 0,065\sqrt{u}; \tag{1.7}$$

at $u > 100 \text{ s}^{-1}$

$$k_u = 0,8 + 0,085\sqrt{u}; \tag{1.8}$$

at $T = 900\text{--}1200\ ^\circ\text{C}$

$$k_t = 0,60 + 0,0045(1200 - T)\sqrt{\frac{1200 - T}{T}}; \tag{1.9}$$

The set of Eqs. (1.4)–(1.9) is a mathematical model of current stress under hot strain of steels investigated by V. Zyuzinym.

In accordance with Eqs. (1.4)–(1.9), the value of the stress current in a hot rolling in any section or at any point of the deformation cell can be calculated if in each of them the values of the parameters ε, T, and u corresponding to this section or given the point.

As a result of the strengthening and influence of the deformation rate, rolling resistance of the deformation of the metal in the cell is rapidly increasing. After the output of the headquarters from the rolls of the stress flow of metal σ_y, rolled in hot state, quickly decreases as a result of abrasion. In the long process of relaxation, the stress flow decreases to σ_{yt}, that is, to the value practically equivalent to the yield strength at a given temperature in the conditions of statistical tests. The above is confirmed by the pictured. 1.9 is the graph of the variation of the deformation resistance σ_s and the yield strength σ_y of the headquarters material between the cell gap of the two adjacent cells of the continuous hot rolling state. From Fig. 1.9 shows that the value of the yield strength at the input of the second cage cell deformation temperature dependent staffs and resistance to deformation at the exit of the first deformation stand and on the duration of transport (relaxation) between the metal cage.

The influence of the duration of relaxation on the stresses of the material flow of the staff during hot rolling can be judged from the graph $\sigma_y = f(\tau)$ (Fig. 1.10) for steel 3 (Standard of Ukraine), built according to Dinnik [15, 16].

From this graph, it can be seen that the most intense stripping of steel 3 (Standard of Ukraine) occurs in the first 0.5–1.5 s after plastic deformation, and after approximately 3–4 s, almost complete deterioration of the metal is observed. When rolling is carried out at high speed, the duration of transport of metal between cages is calculated in tenths and hundredths of a second. In these cases, the metal does not completely absorb and is included in the deformation cell of the next cage with a higher current tension, which leads to an increase in the resistance of the material's deformation of the head when rolling in a given cage. The marked feature of

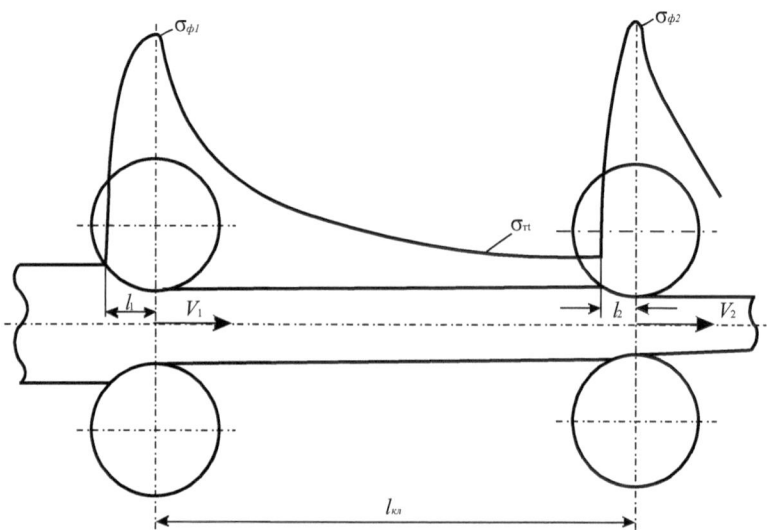

Fig. 1.9 Graph of variation resistance of the deformation resistance and the stresses of the metal head flow during hot rolling in two adjacent cages of a continuous state

Fig. 1.10 Change in the stress of the steel 3 (Standard of Ukraine) flow at $T = 1000\,°C$ during the relaxation process: σ_{s1}—the value of the deformation resistance at the outlet from the deformation cell; σ_{yt} is the value of the current stress at $T = 1000\,°C$

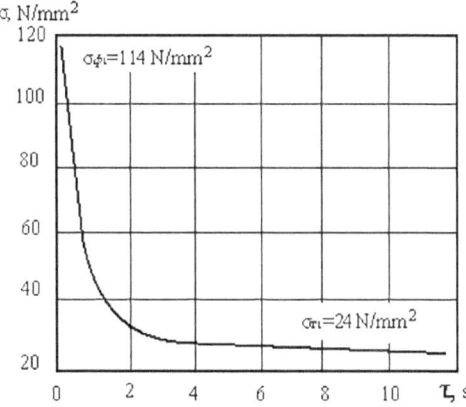

the formation of σ_y and σ_s of the staff of the staff should be taken into account in determining the resistance of deformation at high-speed states.

To calculate the mean value of the resistance of the deformation σ_{sm} during rolling, it is necessary to know the average value of the stress of the flow of the σ_{ym} material of the headquarters in the cell. We find the σ_{ym} value by the method of thermomechanical coefficients. By analogy with (1.4), the mathematical model σ_{ym} for hot rolling is written in the form:

$$\sigma_{ym} = \sigma_0 k_{\varepsilon m} k_{um} k_{tm} \qquad (1.10)$$

$k_{\varepsilon m},\, k_{um},\, k_{tm}$ coefficients taking into account the influence of the average degree, velocity and temperature of deformation during hot rolling on the stresses of flow.

The average relative compression along the length of the deformation cell can be determined from the condition:

$$\varepsilon_m = \frac{1}{l} \int_0^l \frac{h_0 - h_x}{h_0} dx, \qquad (1.11)$$

$$h_x = h_1 + \frac{\Delta h}{l^2} x^2. \qquad (1.12)$$

After the joint solution (1.11) and (1.12) we obtain:

$$\varepsilon_m = \frac{\Delta h}{l h_0} \int_0^l \left(1 - \frac{x^2}{l^2}\right) dx = \frac{2}{3} \frac{\Delta h}{h_0} = \frac{2}{3}\varepsilon. \qquad (1.13)$$

The average temperature of the headquarters in the deformation cell is equal to:

$$t_m = \frac{1}{3}(t_{0n} + 2t_{1n}), \tag{1.14}$$

where t_{0n}, t_{1n} – the temperature of the headquarters at the inlet and outlet of the deformation cell

To find the values of the coefficients $k_{\varepsilon m}$, k_{um}, k_{tm}, use the Eqs. (1.5)–(1.9), substituting them instead of ε, u and and T respectively $k_{\varepsilon m}$, k_{um}, k_{tm}. Further, knowing $k_{\varepsilon m}$, k_{um}, k_{tm} and σ_0 using the model (1.10), we calculate the value σ_{ym}.

Other, more compact models are also known σ_y. These include models:

Zyuzin [13]:

$$\sigma = \sigma_0 \cdot A_1 \cdot A_2 \cdot A_3 \cdot \varepsilon^{m_1} \cdot u^{m_2} \cdot \exp(m_3 T) \tag{1.15}$$

where A_1, A_2, A_3, m_1, m_2, m_3–constant coefficients, which are determined individually for each steel;

Zibel and Pomp [2]:

$$\sigma = \sigma_0 + b \cdot u^m \tag{1.16}$$

where σ_0—yield line at static deformation, kg/mm^2;

b and m are constant coefficients that depend on the nature of the material.

Nadai [10]:

$$\sigma = \sigma_0 + m \cdot \ln \frac{u}{u_0} \tag{1.17}$$

where m—constant coefficient, which depends on the material.

Tselikova and Persiantseva [2]:

$$\sigma = \sigma_0 + D\frac{A}{u}\left(1 - e^{-A\frac{\varepsilon}{u}}\right), \tag{1.18}$$

where

D   reinforcement module, kg/mm^2;

A   is the coefficient of proportionality, which is the relaxation rate, s^{-1};

σ_0 yield line at static deformation, kg/mm^2;

u the average rate of deformation, s^{-1}.

Andreyuk and Tyuleneva [17]:

$$\sigma = s\sigma_0 \cdot (10\varepsilon)^a u^b \left(\frac{T}{1000}\right)^{-c}, \tag{1.19}$$

where

σ_0 the base value of the stress of the material flow headquarters, the corresponding $\varepsilon = 0.1$; and $u = 1\ \mathrm{s}^{-1}$; $T = 1000\ °\mathrm{C}$;

s, a, b, c are constant coefficients determined individually for each steel.

Models of type (1.15)–(1.18) have been proposed for only eight, preferably hardly alloyed steels. In this regard, they have local practical value.

Practical interest is represented by L. Andreyuk – G. Tyulenieva (1.19). Such models have been developed for 66 grades of carbon and alloy steels, including special purpose steels. The model (1.19) is capable of operating in the temperature range of 800–1300°C.

The range of the dependence (1.19) is: T = 800–1300 °C, ε = 5–55%, u = 0.01–150 s^{-1}, experimental data are obtained for a wide range of grades of steels and alloys. According to the authors of the formula (1.19), a possible error in determining the value of the current flow of metal using this formula is 4%. Taking this into account, in the development of an improved model of the hot-rolling process, the formula L. Andreyuk–G. Tyulenieva Significant drawback of L. Andreyuk–G. Tyulenieva is that during tests of most grades of steels the chemical composition and the percentage of the content of alloying elements were regulated by the standards of the 60–80 s of the XX century (GOST 380-60, GOST 1050-88) on the upper limit. In modern production, for the sake of economy, domestic metallurgists tend to minimize the percentage of alloying elements, in addition, there are no reference data for foreign standards (EN10025-93, EN 10130-98, ASTM A366/A366M-91). Thus, the clarification of the value of the current stresses of steels during the hot rolling process, taking into account their actual chemical composition, is another reserve for increasing the accuracy of the power-supply calculation of the large-scale states.

For the determination of the current of the metal during hot rolling, the formula V. Nikolaev is used [18]:

$$\sigma_y = \sigma_{y0} \cdot k_t \cdot k_\varepsilon \cdot k_u, \tag{1.20}$$

where σ_{yo} – the base current voltage is determined by the fixed (basic) values of temperature (T), relative compression (ε) and deformation rate (u):

$$\varepsilon = \frac{\Delta h}{H},$$
$$u = v \cdot \varepsilon / l_d,$$

where k_t, k_ε, k_u,—coefficients taking into account the influence of temperature, relative compression and the rate of deformation.

A mathematical model for calculating the basic stress flow for different grades of steel has the form [19]:

for $N_1 < 5$ (carbon, structural, instrumental, low-alloy steels):

$$\sigma_{y0} = 80 + 25 \cdot \left\{ 1 - \left[\frac{5 - N_1}{4,5} \right]^{1,8} \right\}, \tag{1.21}$$

for $N_2 > 5$ (high-alloy steels):

$$\sigma_{y0} = 110 + 38 \cdot \left\{ 1 - \left[\frac{N_1 - 2}{32} \right]^{0,25 \cdot N_2} \right\}, \qquad (1.22)$$

where N_1 and N_2—the sum of the chemical elements in this steel, % (except iron, sulfur and phosphorus) (Table 1.2).

The formula for the correction factor k_t, depending on the temperature, has the form:

$$k_t = 1.66 - 1.1 \cdot \left(\frac{T}{400} - 2 \right)^{0,7}, \qquad (1.23)$$

where T—temperature of metal, °C.

The calculation of the coefficients is carried out according to the formulas given in Tables 1.3 and 1.4.

In work [20] on the basis of the theory of the planned experiment a scientifically based analysis of the accuracy of known methods for calculating the stress of the flow of metal was performed. The authors developed a computer program that allowed us to evaluate the accuracy of the methods of V. Nikolayev and L. Andreyuk–G. Tyulenyev for 27 structural, instrumental and stainless steels, experimental information for which is given in the work [3]. Average relative error based on V. Nikolaev method is equal to 14.5%, according to L. Andreyuk–G. Tyulenieva–21.2%.

Table 1.2 Steel groups for calculating the stresses of metal flow

Steel group	Metal
Carbon and tool steel	
I	St 50–2; DC 04; C 22; USt 37-2; C 45; C 80W2 (Standard of DIN)
Alloy and low-alloy steels	
II	41Cr4; 100Cr6; 17Mn4 (Standard of DIN); 15CrSiNiCu (Standard of Ukraine), 12ChN3A (Standard of BDS)
III	17CrNiMo6; 105WCr6; 60Si7 (Standard of DIN); 30HGSA (Standard of PN); 18Cr2Ni4MoWN; 35CrNiMn2Mo; 60CrNiMnSi2Mo (Standard of Ukraine)
IV	X8Cr17 (Standard of DIN); 10CrNi (Standard of Ukraine)
V	21NiCrMo2; X12CrMo5 (Standard of DIN); 12CrNiMoVN; 20Cr5NiMn2 (Standard of Ukraine)
VI	20MnCr5G (Standard of DIN); SNC236 (Standard of JIS); 20CrMnNiB (Standard of Ukraine)
VII	X40Cr13; X10CrNiTi18-9; HS18-0-1 (Standard of DIN); Cr16Ni5Mo4; Cr18Ni12Mo2Ti (Standard of Ukraine)

Table 1.3 Formulas for calculating the coefficient k_ε

Steel group	ε	Dependency type
I, II	0, 025...0, 1	$k_\varepsilon = 0, 8 + 0, 2 \cdot \left[1 - 178 \cdot (0, 1 - \varepsilon)^2\right]$
I, II	> 0, 1	$k_\varepsilon = 1 + 0, 43 \cdot \left[1 - 6, 3 \cdot (0, 5 - \varepsilon)^2\right]$
IV	0, 025 ... 0, 1	$k_\varepsilon = 0, 7 + 0, 3 \cdot \left[1 - 50 \cdot (0, 1 - \varepsilon)^2\right]$
IV	> 0, 1	$k_\varepsilon = 1 + 0, 68 \cdot \left[1 - 6, 6 \cdot (0, 5 - \varepsilon)^2\right]$
III, VI, VII	0, 025...0, 1	$k_\varepsilon = 0, 7 + 0, 3 \cdot \left[1 - 50 \cdot (0, 1 - \varepsilon)^2\right]$
III, VI, VII	> 0.1	$k_\varepsilon = 1 + 0, 38 \cdot \left[1 - 6, 3 \cdot (0, 5 - \varepsilon)^2\right]$
V	0, 025...0, 1	$k_\varepsilon = 0, 8 + 0, 2 \cdot \left[1 - 178 \cdot (0, 1 - \varepsilon)^2\right]$
V	> 0, 1	$k_\varepsilon = 1 + 0, 5 \cdot \left[1 - 9 \cdot (0, 5 - \varepsilon)^2\right]$

Table 1.4 Formulas for calculating the correction factor k_u

Borders change u, s^{-1}	The formula for k_u	Steel group
0, 4...10 0, 001...0, 4	$k_u = 0, 38 + 0, 065 \cdot (7 + \ln u)$ $k_u = 0, 22 + 0, 072 \cdot (7 + \ln u)$	I, II
0, 001...10	$k_u = 0, 38 + 0, 065 \cdot (7 + \ln u)$	III – VII
≥ 10	$k_u = 1, 03 + 0, 1 \cdot ((\ln u) - 2, 3)^{1,5}$ $k_u = 1, 03 + 0, 11 \cdot ((\ln u) - 2, 3)^{1,05}$ $k_u = 1, 03 + 0, 13 \cdot ((\ln u) - 2, 3)^{2,2}$	I, II III–V, VII VI

1.5 Perspective Technological Schemes of Hardening

In modern conditions, the main task in the production of rolled metal is to improve the quality of metal products, as well as reduce waste during its production and processing [21]. In implementing the program of saving ferrous metals an important place is occupied by the introduction of thermomechanical processing technologies that improve the mechanical properties of rolled metal. At the country's metallurgical plants, thermomechanical processing of rolled products in thermal furnaces is very common, and for many types of profiles it is mandatory. The operation of accelerated cooling of rolled products is a thermomechanical treatment that allows to eliminate or reduce heat treatment of metal from special heating in thermal furnaces. Therefore, this operation becomes important [22].

As is known, the main requirement for low alloy structural steels is to obtain a sufficiently high level of mechanical properties, which are determined by its structural state. The structural state of steel, mainly depends on the modes of thermomechanical processing and is determined by the chemical composition [23–26].

Thermomechanical treatment (TMT) rolled is the most effective way to increase its mechanical properties, as a result of which the structure of steel is formed. Formation of the final structure of steel occurs in conditions of increased density of imperfections in the structure created by plastic deformation.

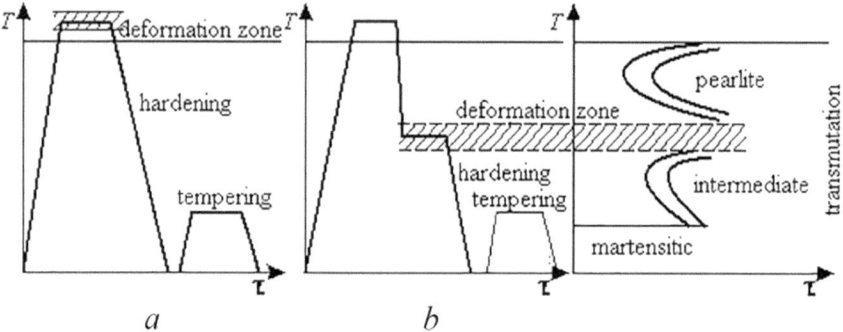

Fig. 1.11 Diagram of modes of thermomechanical processing [25]: **a** high-temperature thermo-mechanical treatment; **b** low-temperature thermomechanical treatment

The advantage of thermomechanical processing is that with a significant increase in strength, the ductility characteristics are reduced slightly, and the toughness is 1.5 to 2 times higher than the toughness for the same steel after quenching with low tempering.

At present, several types of TMT are known: high-temperature, low-temperature, mechanical-thermal, preliminary thermomechanical treatment, and others. The recently gained controllable rolling (CR) is also considered a variant of TMT [24].

The essence of high-temperature thermomechanical treatment consists in heating the steel to the austenitic state (above A3). At this temperature, steel is deformed, which leads to a hardening of the austenite. Steel with such a state of austenite is quenched (Fig. 1.11a). High-temperature thermomechanical treatment practically eliminates the development of temper brittleness in a dangerous temperature range, weakens irreversible release brittleness, and sharply increases the toughness at room temperature. The temperature threshold of cold shortness is lowered. High-temperature thermomechanical treatment increases the resistance to brittle fracture, reduces the sensitivity to cracking during heat treatment.

High-temperature thermomechanical treatment is efficiently used for carbon, alloyed, structural, spring and tool steels. Subsequent tempering at a temperature of 100–200 °C is carried out to maintain high strength values.

At low-temperature thermomechanical processing, the steel is heated to an austenitic state. Then it is kept at high temperature, cooled down to a transformation temperature (400–600 °C), but below the recrystallization temperature, and pressure treatment is performed at this temperature (Fig. 1.11b). Low-temperature thermomechanical treatment, although it gives higher hardening, does not reduce the tendency of steel to temper brittleness. In addition, it requires high degrees of deformation, therefore, it requires powerful equipment.

Low-temperature thermomechanical treatment is applied to medium-carbon alloyed steels that have a secondary stability of austenite.

The increase in strength in thermomechanical processing is explained by the fact that as a result of the deformation of austenite, its grains (blocks) are crushed.

Dimensions of blocks are reduced by two to four times compared to conventional hardening. The dislocation density also increases. Upon subsequent quenching of such austenite, smaller plates of martensite are formed, and stresses decrease.

Low-temperature processing has received little application [27]. The most frequently used high-temperature treatment. Its advantage is that the workpieces immediately after the end of the hot deformation processing can be subjected to quenching without special heating, using only the heat of the metal after hot deformation. Thus, fuel economy is achieved, the time for obtaining finished products is reduced, mechanical properties are increased, and the strength, plasticity, toughness and heat resistance of steel are increased [24, 27].

Controllable rolling is one of the promising types of thermomechanical processing of low-alloy steels and represents hot rolling according to a predetermined regime, including programmed temperatures of the beginning and the end of deformation, reduction and cooling rate at various stages of plastic processing [27]. With controlled rolling, by reducing the deformation temperature in the accelerated inter-cell cooling units in combination with accelerated cooling of the finished rolled steel, the structure of the steel is formed with a fine grain of ferrite, as the yield strength increases, the temperature decreases and weldability improves [25]. Controllable rolling allows you to obtain a pearlite grain with a diameter of 5...10 μm or less, which leads to steel control by 10...30% while maintaining high ductility and viscosity [24].

The results of the studies [28, 29] showed that the effect of controlled rolling is associated not only with grain refinement, but also with the creation of a stable substructure, and in many cases the influence of the substructure is predominant.

Y. Matrosov singled out eight stages of controlled sheet rolling technology [30–35]:

(1) Austenization at temperatures ensuring a sufficiently uniform structure of the metal before rolling. The rolling temperature for most steels, micro-alloyed with niobium, vanadium and titanium, is about 1150...1200 °C;

(2) High-temperature deformation of stable austenite in the region of rapidly occurring recrystallization processes, when the deformation temperature is higher than the recrystallization temperature. The purpose of high-temperature deformation is to obtain as small a grain of austenite as possible by alternating multiple reduction and recrystallization. For low-alloyed steel with niobium, the degree of deformation for the development of dynamic recrystallization at 1100–1150 °C is 40–60%. However, the implementation of such modes in industrial conditions is difficult. Rough rolling on the existing plate mill is carried out at temperatures not lower than 980...1000 °C with compression ratios per pass 15–20% (Fig. 1.12b);

(3) The average temperature deformation in the lower part of the austenite region is carried out with the aim of increasing the density of defects in the crystal structure of the metal and their ordered distribution (substructure), which leads to multiple formation of ferritic volumes in a polymorphic austenite-perlite transformation;

(4) Deformation of austenite in the region of polymorphic γ-α transformation;

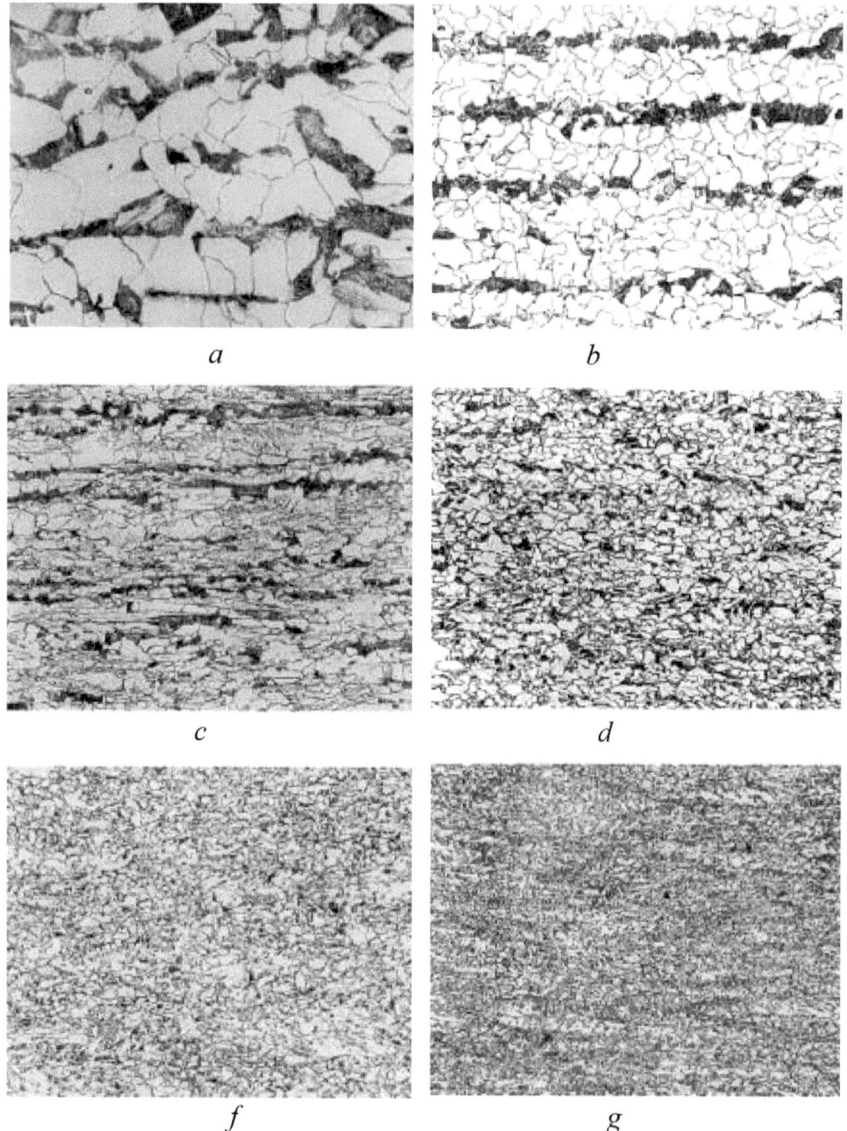

Fig. 1.12 The effect of TMT and chemical composition on the structure of steels [36, 37]: **a** hot rolling; **b** CR in the austenitic region; **c** CR in the ($\gamma + \alpha$) region; **d–g** various options for accelerated cooling and alloying of steel: ferrite + perlite; quasi-polygonal ferrite; bainite, × 500

(5) Deformation in the two-phase γ-α region. The lowering of the deformation temperature in the γ-α region promotes hardening of the steel, since in this case the fraction of ferrite grains hardened by deformation increases (Fig. 1.12c);

(6) Deformation in a three-phase region, which is advisable to carry out, if in the complex of mechanical properties, the paramount importance is attached to obtaining a very high strength;

(7) The deformation below the A_{r1} point is possible with the availability of powerful rolling equipment and with low requirements to the plastic characteristics of rolled products:

(8) The cooling of the steel after the deformation is completed is carried out in the air at a cooling rate of 0.5–1 °C/s or in laminar cooling of rolled products at a rate of about 15 °C/s (Fig. 1.12d).

In foreign and domestic practice under industrial conditions, controlled sheet rolling is carried out in two or three stages (Fig. 1.13). The first stage starts from the heating temperatures for rolling up to 950–1000 °C, when austenite recrystallizes rapidly, which ends in the interdeformation of the inter-information pause. In the second stage of the process, recrystallization in conventional steels becomes more difficult in the temperature range from 950 °C to A_{r3}, while in steels with niobium it is practically suppressed. In this region, the temperature and the degree of deformation have a significant effect on the kinetics of recrystallization.

If the temperature-deformation conditions are chosen correctly, the grain decreases as a result of static recrystallization. Deformation leads to the formation of slip bands and the release of dispersed phases. With an increase in the degree of deformation, the number of slip bands increases and the uniformity of their location increases, which facilitates obtaining a fine uniform grain of ferrite after conversion. As a result, the cold resistance of the steel improves.

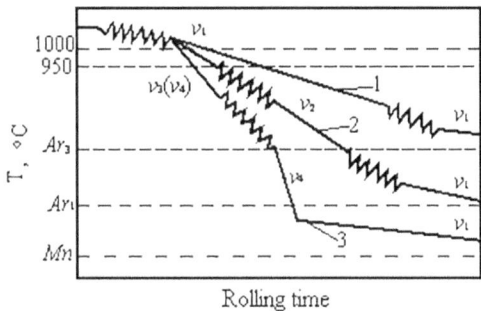

Rolling time

Fig. 1.13 Technological schemes of controlled rolling with various cooling methods, devices with chambers of complex profile [36, 38]: 1, 2—rolling a sheet of steel with Nb and V respectively in two and three stages; 3 - rolling sheet from steel without microadditives; $v_1 = 3$–7 °C/s, cooling (grooming) in the air; $v_2 = 5$–25 °C/s, odor installations; $v_3 = 50$–70 °C/s, water–air mixture; $v_4 = 150$–200 °C/s and above

In the third stage, at temperatures below Ar3, dispersion hardening processes with grain refinement are accompanied by development of grain subgrains, the latter two factors being crucial in improving the properties of the steel.

The authors of Ref. [29] indicate four factors that determine the grinding of ferritic grain under controlled rolling. This, first, a reduction in the temperature of heating for rolling, leading to a decrease in the size of the initial, and consequently, recrystallized austenite grain. The second factor is the retardation of recrystallization during hot deformation, which can be achieved in various ways: by lowering the temperature and increasing the degree of deformation, by raising the recrystallization temperature by doping the solid solution, separating out of the solid solution before recrystallization or during its second phase dispersed particles that inhibit the migration of the boundaries grains and blocks. The third factor is a decrease in the temperature of the γ-α transformation, which can be achieved both by appropriate doping and by controlling the rate of post-deformation cooling. Finally, an important factor is the prevention of the growth of ferritic grains in the upper part of the ferrite region, for example, when the steel is cooled in rolls. These four mechanisms do not exhaust all possibilities of grinding ferrite grain under controlled rolling. A fine ferrite grain can also be obtained from a non-recrystallized or partially recrystallized austenite grain with a high defect density of the crystalline structure as a result of ferrite firing on the defects within the austenite grain.

It was shown in [32] that controlled rolling of steel 09Mn2 (Standard of Ukraine) with an end in a dual-phase region at 750–700 °C allows, together with higher strength properties, to increase the impact toughness at low temperatures by 300–400 kJ/m^2, and also by about 40 °C to lower the critical temperature of brittleness as compared with the end of rolling in the lower part of the γ-region. In the case of the end of rolling at 700 °C, the following values of the mechanical properties determined during the tensile test were obtained: yield strength 450 MPa, temporary tensile strength 555 MPa, elongation about 27%. Further reduction of the deformation temperature provides an increase in strength characteristics with a noticeable decrease in ductility and toughness at room temperature. The author of [33] relates the effect of increasing mechanical characteristics to the refinement of ferritic grains and the creation of a stable dislocation substructure in a ferrite.

Thus, the available information shows the principal possibility of increasing the level of mechanical properties of rolled products from dual-alloy steels. As a result, it becomes possible to replace hard-alloy steels. Thus, in work [35] it was shown that in the case of completion of its rolling at low temperatures, 09Mn$_2$ (Standard of Ukraine) steel acquires a complex of properties similar to that obtained for microalloyed steel 08Mn$_2$VMoNb (Standard of Ukraine). However, the need to lower the rolling temperature [39] to intercritical or even subcritical values creates certain difficulties in realizing the process on the existing rolling equipment [40].

References

1. Vasylev Ya. ta Minaiev O (2009) Teoriia pozdovzhnoi prokatky. UNITEKh, Donetsk, p 488. ISBN 978-966-525-968-8 (in Ukrainian)
2. Tafel W (1931) The theory and practice of rolling steel. The Penton Publishing Co., Cleveland, O
3. Ginzburg VB (1989) Steel-rolling technology: theory and practice. Taylor & Francis Group, 791 p
4. Ginzburg VB (1993) High-quality steel rolling: theory and practice. Taylor & Francis Group, 832 p
5. Kurrein M (1964) Plasticity of metals: The mechanical behavior and the changes in structure of metals under plastic deformation. Griffin, London, p 270
6. Chekmarev AP, Rydner ZA (1957) Prokatnoe proyzvodstvo. T. 2, 2. Yzd. AN USSR; 1957, Kyev (in Russian)
7. Laasraoui A, Jonas J (1991) Prediction of steel flow stresses at high temperature and strain rates. Metall Trans A 22:1545–1558
8. Moreira A, Junior J, Balancin O (2005) Prediction of steel flow stresses under hot working conditions. Mater Res 8(3):309–315
9. French HJ (1922) Effect of heat treatment on the mechanical properties of 1% carbon steel. Government Printing Office, Washington, p 121 c
10. Felbeck (1995) Strength fracture engineering solids. Longman Higher Education Division (a Pearson Education Company)
11. Cook PM (1957) The real curves, stress rate of deformation for the steels by reduction. The Institution of Mechanion Engineer, pp 75–77
12. Hoogendoorn TM, van den Hoogen AJ (1994) Heat treatment in hot strip rolling of steel. Mater Sci Forum 163–165:51–62. https://doi.org/10.4028/www.scientific.net/msf.163-165.51
13. Ziuzyn VY (1963) Opredelenye soprotyvlenyia deformatsyy metodom termomekhany-cheskykh koeffytsyentov. Trudi VNYYMETMASh 8:74–89 (in Russian)
14. Klymenko P (2011) Uprochnenye staly y tsvetnikh metallov pry kholodnoi y horiachei deformatsyy. Porohy, Dnepropetrovsk (in Russian)
15. Kubotera H, Nakaoka K, Nagamine T (1966) Hot rolling texture of low carbon steel. Tetsu to Hagane 52(8):1171–1179. https://doi.org/10.2355/tetsutohagane1955.52.8_1171
16. Dynnyk A (1962) Ystynnie predeli tekuchesty staly pry horiachei prokatke. Teoryia prokatky. 157–72 (in Russian)
17. Andreiuk L, Tiulenev H, Prytsker B (1972) Analytycheskaia zavysymost soprotyvlenyia deformatsyy stalei y splavov ot khymycheskoho sostava. Stal 6:522–523 (in Russian)
18. Nykolaiev VO, Mazur VL (2010) Vyrobnytstvo ploskoho prokatu. ZDIA, Zaporizhzhia (in Ukrainian)
19. Tsyganov V, Sheyko S (2022) Features of engineering the wear-resistant surface of parts with the multicomponent dynamic load. Wear 494–495:204255. https://doi.org/10.1016/j.wear.2022.204255
20. Yakovchenko OV, Pugach OA, Ivleva NI (2011) The Analysis of precision of the existing methods of evaluation of metal flow tension, depending on steel chemical composition. Bull Priazov State Tech Univ 2(23):69–80
21. Matsudo K, Shimomura T, Osawa K (1980) Production of high strength gold rolled steel sheets by the NKK continuous annealing line process. Fachberichte Hutten-praxis Metallweiter verarbeitung 12:1128–1136
22. Fujibayashi A, Omata K (2005) Steel's advanced manufacturing technologies for high performance steel plates. JFE Tech Rep 5:68–72
23. Brovkin V (2007) Yssledovanye teplovykh protsessov kontrolyruemoi prokatky na stane 250. Metallurhycheskaia y hornorudnaia promyshlennost 3:110–114 (in Russian)
24. Uzlov YH, Savenkov VIa, Poliakov SN (1981) Termycheskaia obrabotka prokata. Tekhnyka, Kyiv (in Russian)

25. Bolshakov V, Dolzhenkov Y, Dolzhenkov V (2002) Termycheskaia obrabotka staly y
    metalloprokata. Sich, Dnepropetrovsk (in Russian)
26. Belodedenko S, Grechany A, Yatsuba A (2018) Prediction of operability of the plate rolling
    rolls based on the mixed fracture mechanism. Eastern-Eur J Enterprise Techn 1, 7(91): 4–11.
    https://doi.org/10.15587/1729-4061.2018.122818
27. Kuznetsov YuV, Brovkin VL, Ivanova GN, Duduka VA (1991) Improving of rolled products
    quality during deformation at low heating temperatures. Stal' 11:65–67
28. Bolshakov VI, Rychahov VN, Frolov VK (1994) Termycheskaia y termomekhanycheskaia
    obrabotka stroytelnykh stalei. Sich, Dnepropetrovsk (1994) (in Russian)
29. Bernshtein ML (1977) Struktura deformyrovannykh metallov. M.: Metallurhyia (in Russian)
30. Pohorzhelskyi VY, Lytvynenko DA, Matrosov YuY, Yvanytskyi AV (1979) Kontrolyruemaia
    prokatka. M.: Metallurhyia (in Russian)
31. Sheyko S, Tsyganov V, Hrechanyi O, Vasilchenko T, Hrechana A (2024) Determination of the
    optimal temperature regime of plastic deformation of micro alloyed automobile wheel steels.
    Res Eng Struct Mater 10(1): 31–339. https://doi.org/10.17515/resm2023.49me0428tn
32. Matrosov YuY (1987) Kontrolyruemaia prokatka – mnohostadyinyi protsess TMO nyzkolehy-
    rovannykh stalei. Stal 7:75–80 (in Russian)
33. Matrosov YuI, Fylymonov VN, Holovanenko SA (1979) Uluchshenye mekhanycheskykh
    svoistv malouhlerodystoi staly 09G2. Chernaia metallurhyia Biulleten NTY. (14):39–41 (in
    Russian)
34. Pohorzhelskyi VI (1986) Kontrolyruemaia prokatka neprerivnolytoho metalla. M.: Metal-
    lurhyia (in Russian)
35. Matrosov YuI, Fylymonov VN, Bernshtein ML (1979) Vlyyanye drobnoi deformatsyy v γ+α y
    α-oblastiakh na mekhanycheskye svoistva staly 09G2. Yzvestyia vuzov Chernaia metallurhyia.
    (11):115–119
36. Minaev AA (2008) Sovmeshchennyye metallurgicheskiye protsessy: monograph. Technopark
    DonSTU UNITECH, Donetsk, 552 p (in Russian)
37. Baranov AA, Minaev AA, Geller AL, Gorbatenko VP (1987) Combined working and heat
    treatment of steel. Sov Mater Sci Rev 1(1):33–36
38. Baranov AA, Minaev AA, Gorbatenko VP, Demidovich EA, Cherednichenko AL (1983) Influ-
    ence of conditions of cooling after controlled rolling on structure and properties of medium
    carbon steels. Steel in the USSR 13(12):564–566
39. Belodedenko SV, Hanush VI, Hrechanyi OM (2022) Experimental verification of the surviv-
    ability model under mixed I+II mode fracture for steels of rolling rolls. Struct Integr 25:3–12.
    https://doi.org/10.1007/978-3-030-91847-7_1
40. Belodedenko S, Hanush V, Hrechanyi O (2022) Fatigue lifetime model under a complex loading
    with application of the amalgamating safety indices rule. Procedia Struct Integr 36:182–189.
    https://doi.org/10.1016/j.prostr.2022.01.022

Chapter 2
Modelling of the Process of Plastic Shaping and Its Influence on the Structure of Dual-Phase Steels

Dual Phase (DP) steels with a ferritic-martensitic or ferritic-perlite structure have high strength properties. "Soft" ferrite (up to 80%) imparts high plastic properties to DP steels in the initial state. During the stamping process, deformation stresses are concentrated in the ferrite phase, and a high degree of strain hardening (combined with high elongation) is achieved, which guarantees a very high tensile strength of DP steels [1]. Compared with low-alloy structural high-strength steels (HSLA) having the same yield stress, DP steels show a higher initial strain hardening rate, higher elongation and tensile strength, and a lower σ_s/σ_v ratio. The magnitude of the time resistance of DP steels reaches 1000 N/mm^2 (DP 700/1000). In DP steels, carbon provides the formation of the martensitic phase and, in combination with the balanced additives Mn, Cr, Mo, V and Ni, their strength properties. The composition of the two-phase steels is very diverse, for example, the composition of hot-rolled steel is directly related to the technological capabilities of the equipment: the greater the cooling capacity on the outfitting roller table of the mill and the lower possible quenching temperature, the lower the content of alloying elements. The basic scheme for obtaining a two-phase structure-the allocation of the necessary amount of ferrite and subsequent intensive cooling to obtain martensite-is shown in Fig. 2.1 [2].

The high ability to strain hardening causes a good redistribution of stresses and, consequently, stamping. The yield point of the finished part is significantly higher than the original workpiece. High final mechanical properties provide high fatigue strength and high ability to absorb energy, enabling to use them in structural elements and fastening elements. However, very high-strength metal is required for the manufacture of car wheel parts. As a result, their deformation during production is not sufficient to obtain the advantages of two-phase steel. For this, two-phase steels of a wide range of strengths have been developed: DP 450, 500, 600, 780, 980, 1180 with increased deformability. Here the main idea is to increase the strength with increasing the volume fraction of martensite. The steels are produced in cold-rolled and hot-rolled (DP 600) states [3, 4].

© The Author(s), under exclusive license to Springer Nature Switzerland AG 2024
S. Sheyko et al., *Thermoplastic Processing of Structural Metallic Materials*,
SpringerBriefs in Materials, https://doi.org/10.1007/978-3-031-73896-8_2

Fig. 2.1 Scheme of
obtaining a dual-phase
structure of steel on a
continuous broadband mill
[1]

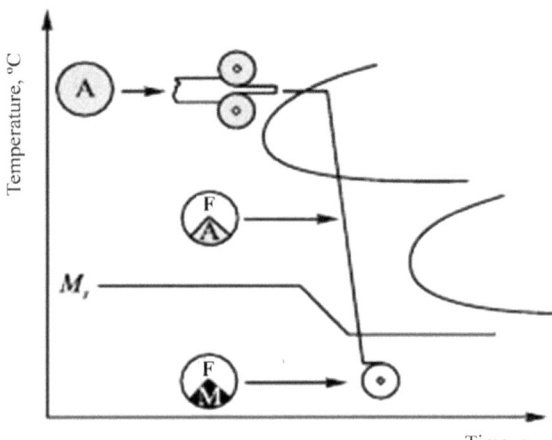

2.1 Theoretical Features of the Investigation of the Influence of the Stress–Strain State on Structural Transformations in Dual-Phase Steel

On the basis of theoretical and experimental studies, we shall determine the integral characteristics of the stress state at a point, taking into account the Huber-Mises model, establish the mechanical characteristics of the plastic medium (yield stress), variants of deformation, temperature, and velocity loading. From the proposed values of thermomechanical parameters, we choose from the recrystallization diagram for this steel grade those that correspond to the required structural state. This requires carrying out not only theoretical studies, but also experimental ones that allow us to reveal the patterns of changes in mechanical characteristics, physical properties of the thermomechanical factors of the process, and the chemical composition of the steel. We determine the influence of thermomechanical parameters on the structural state of a plastic medium for a particular grade of steel.

The inhomogeneity of the deformed state is the main feature of metal products, formed by pressure processes. As a consequence, the metal of the product in its volume has different mechanical properties and fatigue resistance. This is all the more important because, in each model, for individual characteristics, as a rule, individual elements or zones are responsible. Therefore, in forecasting, it is necessary to use not the averaged strain characteristics, but to determine the deformation of material points of the entire product volume. As an index, use the shear stress intensity:

$$\tau_i = \frac{1}{\sqrt{6}}\sqrt{\left(\sigma_x - \sigma_y\right)^2 + \left(\sigma_y - \sigma_z\right)^2 + (\sigma_z - \sigma_x)^2 + 6\left(\tau_{xy}^2 + \tau_{yz}^2 + \tau_{zx}^2\right)}. \quad (2.1)$$

Real technological processes are characterized by a multistage nonmonotonic nature of the intensity of deformation with complex loading. In the theory of small deformations, it is proved that the deformed state of any material point is completely determined by six components: the three main components, the intensity and the type of deformation. In this case, the total shear deformation is a quantitative characteristic of the degree of change in the shape of the particle under consideration, which is expressed in terms of the principal components of the deformation:

$$\Gamma_i = \sqrt{\frac{2}{3}}\sqrt{\left(\varepsilon_x - \varepsilon_y\right)^2 + \left(\varepsilon_y - \varepsilon_z\right)^2 + \left(\varepsilon_z - \varepsilon_x\right)^2 + \frac{3}{2}\left(\gamma^2 + \gamma\right)} \qquad (2.2)$$

In the general case, for a finite (significant) deformation, the intensity expression is conserved. For multioperational (multi-transient) processes Smirnov-Alyaev proposed a mathematical expression according to which the total (resulting) degree of deformation for the entire technological process is defined as the arithmetic sum of the deformation degrees of individual operations, whose values, in the case of a monotonous deformation process, are numerically equal to the intensity of the main deformations. It was noted that the intensity is a scalar quantity and the only comparable characteristic of the shape change, which makes it possible to determine the work expended, and the mechanical properties of the deformed body [5].

Approximate theoretical calculations based on an estimate of the magnitude of the amplitude of the change in the potential energy of the nucleus during dislocation motion have shown that the minimum tangential stress necessary for dislocation motion is:

$$\tau_n = \frac{2G}{k} e^{-\frac{2\pi}{k}\frac{a}{b}}, \qquad (2.3)$$

where a is the distance between adjacent slip planes.

It follows from this formula that the larger a and less than b, the smaller τ_n. It is known that the value of a is maximal for close-packed atomic planes, and the smallest value of b corresponds to the most closely packed directions.

Thus, checking condition:

$$\tau_i > \tau_n, \qquad (2.4)$$

one can speak of the possibility of motion of dislocations. Especially mobile are dislocations, providing a plastic shear along the directions and planes most densely populated with atoms. By preventing in any way sliding in these planes, one can call it in those planes where the packing of atoms is less dense.

The necessity of an analytical solution of the spatial problem of the theory of plasticity is obvious. In general, the strained and deformed state of the metal is different at each point of the deformation focus. This leads to heterogeneity of physical and mechanical properties of the metal, ambiguity in determining the power parameters of the process, energy consumption. There appeared works showing the effect of plastic deformation on structural-phase transformations in a metal. In this connection, the determination of the stress state at each point of the deformation center is an actual problem.

In general, the strength of metals increases with decreasing grain size. It is known that this dependence follows the well-known Petch-Hall relation [6–8]:

$$\sigma_y = \sigma_i + k \cdot d^{-\frac{1}{2}}, \qquad (2.5)$$

where

σ_y is the tensile yield strength or the flow stress;
d is the grain size;
σ_i and k are the parameters characterizing this material

Compared with unidirectional static loading, when after penetration by the propagating current of the first large-angle grain boundary, until the final destruction, the metal withstands considerable plastic deformation, under cyclic loading, the picture is much more complicated. Already at the initial stages of alternating deformation, displacement of dislocations is limited by the volume of the grain of ferrite of low-alloyed two-phase steel, and the state of the solid solution necessarily influences the process of formation of the microcrack nucleation center. The transition from the incubation period of growth of the microcrack to the accelerated one, in the literature, is associated with the intersection of the first large-angle grain boundary [9]. Starting from this point, the growth of the fatigue crack becomes even more dependent on the grain size of the ferrite.

As a result of the tests of the steel with different grain size, the calculation of the stress σ_i from the coefficient of friction and the shape factor, the dependence σ_T on d was constructed, which is shown in Fig. 2.2.

It is established that the change in σ_y from the grain size of the ferrite d (Fig. 2.2) obeys the relation (2.5) For ferrite, in the absence of a carbide phase in the steel structure, the value $\sigma_i = 15$ MPa, which fits perfectly into the range of values 8–17 MPa [10] From the construction of the ratio σ_y to d (for static loading conditions of the steel under study), the constants (2.5) were: $\sigma_i = 50$ MPa and $k = 10$ N/mm$^{3/2}$.

In the Petch Hall model, the size of the grain also affects the number of dislocations n accumulating on a length L between the dislocation source inside the grain and the boundary, which in turn affects the concentration of the stress τ_q in the cluster head dislocation, which causes the multiplication or movement of dislocations at point A, just before the cluster, or at point B, spaced at a distance l from it. According to Cottrell, the stress concentration at point A is determined by expression:

$$\tau_q(A) = n(\tau - \tau_i), \qquad (2.6)$$

where τ is the applied shear stress;

τ_i is the frictional stress in the slip plane, which prevents the movement of dislocations.

Fig. 2.2 Dependence of the yield strength σ_y on the grain size d: $d - 1$—250 μm, $d - 2$—200 μm, $d - 3$—150 μm, $d - 4$—100 μm, $d - 5$—50 μm, $d - 6$—20 μm

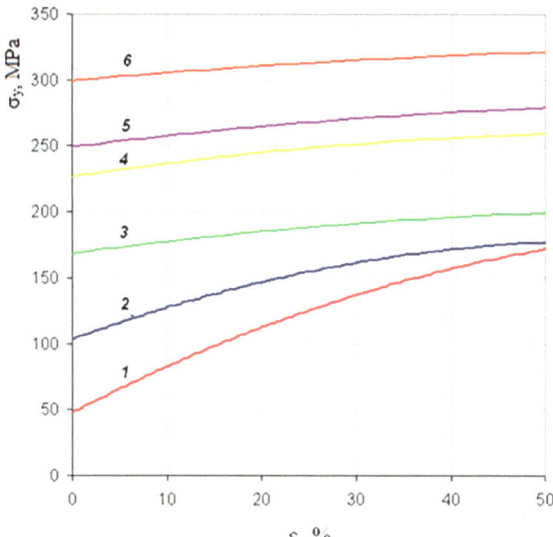

The number of dislocations in the cluster is:

$$n \approx \frac{2L}{b}\left(\frac{\tau - \tau_i}{\mu}\right), \tag{2.7}$$

where b is the value of the Burgers vector;

μ is the shear modulus.

On the other hand, the number of movable dislocations ρ required to maintain the flow of the metal can be estimated by the relationship [10, 11]:

$$\rho = \left(\frac{\sigma_d}{\alpha\mu b}\right)^2, \tag{2.8}$$

where σ_d—stress required for the generation of ρ mobile dislocations; α—coefficient that takes values from 0.1 to 1.0; μ—shear modulus; b—magnitude of the Burgers vector.

After substituting in (2.8) the Huber-Mises condition (4.44) instead σ_d of the value σ_i, $\alpha = 0.6$ (the average value of the interval is 0.1–1.0), $\mu = 8.2 \cdot 10^4$ MPa and $b = 2.48 \cdot 10^{-7}$ mm (for ferrite), the values of the number of mobile dislocations ρ were calculated. The nature of the dependence ρ on grain size d, for the conditions of cyclic loading of the steel under study, is shown in Fig. 2.3

A comparative analysis ρ with a similar characteristic of the steel under study, for static expansion conditions (ρ_1), showed that as the grain size of the ferrite increases, the difference between them decreases. Thus, for grain sizes of ferrite 15–20 μm, $\rho = 7.7 \cdot 10^6$ mm^{-2}, $\rho_1 = 7.0 \cdot 10^6$ mm^{-2}; for $d = 110$–120 μm, $\rho = 4.3 \cdot 10^5$ mm^{-2}, $\rho_1 = 4.25 \cdot 10^5$ mm^{-2}.

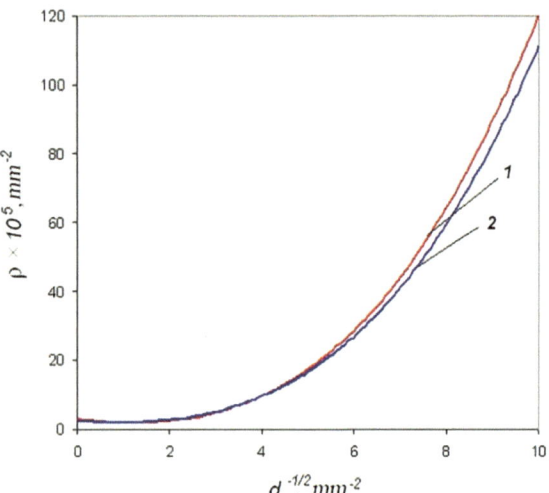

Fig. 2.3 Dependences of the number of mobile dislocations ρ on the grain size of ferrite d of low-alloy steel: 1—Conditions of cyclic loading; 2—Static stretching conditions

From the results obtained, it follows that with increasing grain d, the significance of this indicator, in the level of strength of steel, is reduced. For grain sizes of more than 100 μm, the main effect on the yield stress σ_y is exerted by the stress intensity σ_i, as a result of the transition to solid solution hardening.

Thus, an experimental-theoretical method has been developed for determining the relationship between the stress–strain state of a metal, the grain size d, and the yield point σ_y. To obtain a complete picture of the effect of plastic deformation on the changes in structural-phase transformations in the new steel grade 10HFTBch (Standard of Ukraine), more complete studies are required with the use of modern plastometric methods for testing the stress–strain flow of a metal and its mathematical description.

To evaluate the possibility of using the developed methodology, the calculations of the analytical solution were compared with the results of finite element modeling in the Deform-3D program for thin and thick bands (Fig. 2.4).

For modeling, a rigid-plastic medium and rolled steel—steel 15 (Standard of Ukraine) were used. The contact friction condition was set according to Amanton-Coulomb. For the roll and rolls, grids with a different number of elements (25,000 and 10,000 elements, respectively) were generated. The average value of the yield point in the deformation center was taken as the average value of stress intensity, which was found by integration over the longitudinal section of the focus. The predicted value of the yield stress σ_τ was calculated from formula (2.5).

A comparison of the yield stress values calculated from formula (2.5) and the values found by finite element modeling is shown in Fig. 2.5. The value of the reliability index of the approximation $R^2 = 0.9514$, shows that the degree of coincidence of the results of calculations and simulation results is 95.14%.

Thus, it can be concluded that the developed technique, by definition of the yield point, shows more accurate results in a wide range of thermomechanical parameters.

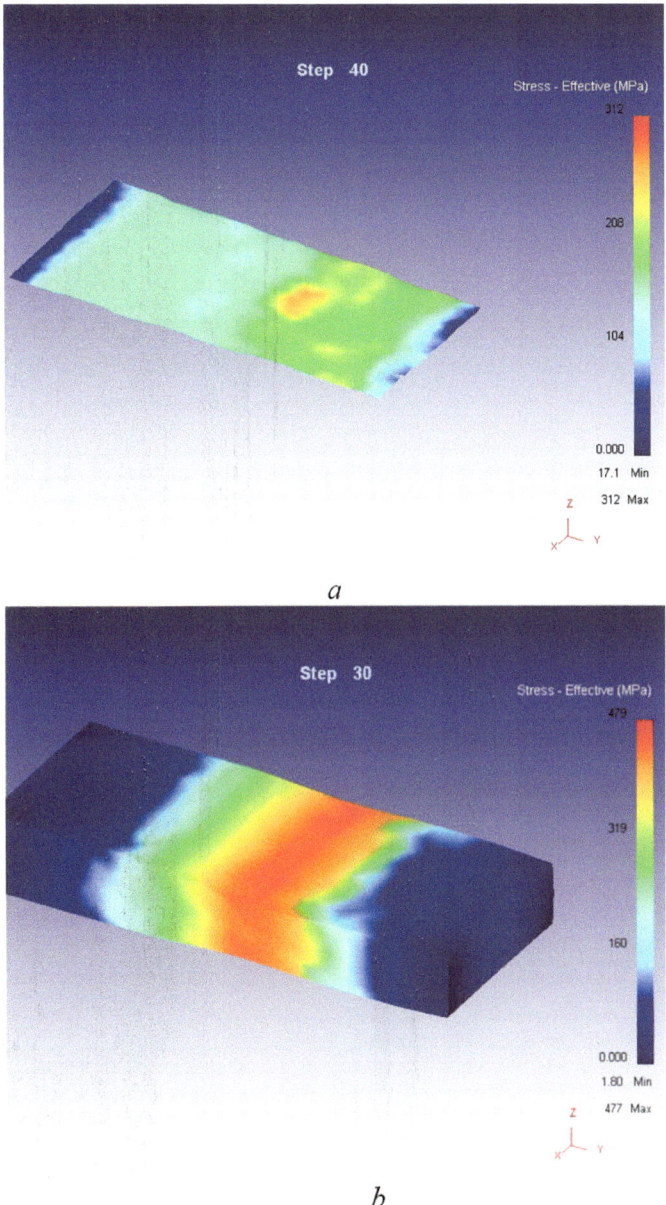

Fig. 2.4 Results of stress intensity modeling in the Deform-3D program for rolling: **a** thin strips, **b** thick bands

Fig. 2.5 Comparison of the yield stress σ_y, calculated by the developed method with the values of σ_i obtained in Deform-3D

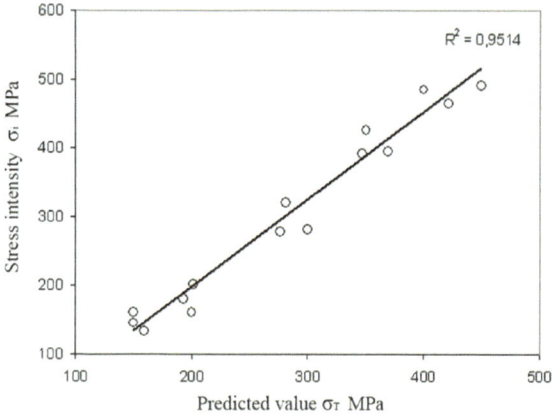

2.2   Experimental Studies of the Effect of Thermoplastic Parameters of the Deformation Process on the Structural State of Dual-Phase Steels

Experimental studies of the modes of thermoplastic deformation were carried out on the Gleeble-3800. The Gleeble-3800 complex (Fig. 2.6) is made in a modular design and is designed to simulate the processes of thermomechanical processing. To the main power unit modules are connected to perform various tasks:

- deformation by compression or stretching;

Fig. 2.6 The Gleeble System 3800

- torsion;
- impact deformation;
- multiaxial deformation.

To enable a wide range of studies, the Gleeble System 3800 has the following main technical characteristics [12]:

- maximum compression force—200 kN;
- maximum tensile force is 100 kN;
- programmable speed of traverse movement—up to 2500 mm/s;
- maximum temperature is 1750 °C;
- heating rate—up to 10,000 °C/s;
- cooling rate—up to 10,000 °C/s.

The Gleeble System 3800 is made in a modular design, which allows you to flexibly change its configuration, depending on the needs of the researcher. In addition, the design of the complex is sufficiently "open" and allows the use of various additional devices in the processing of materials. The working block, Fig. 2.7, was developed to simulate the shock loading of the samples (Fig. 2.8), with high accuracy in degree and strain rate. The special design of the module allows performing multi-stage consecutive loading of fully reproducing deformation modes in virtually any modern rolling mill.

Form of recording in HDS format. This format for describing the test program is designed specifically for working with the Hydrawedge module. This recording format allows the programming of process parameters in a format simulating the

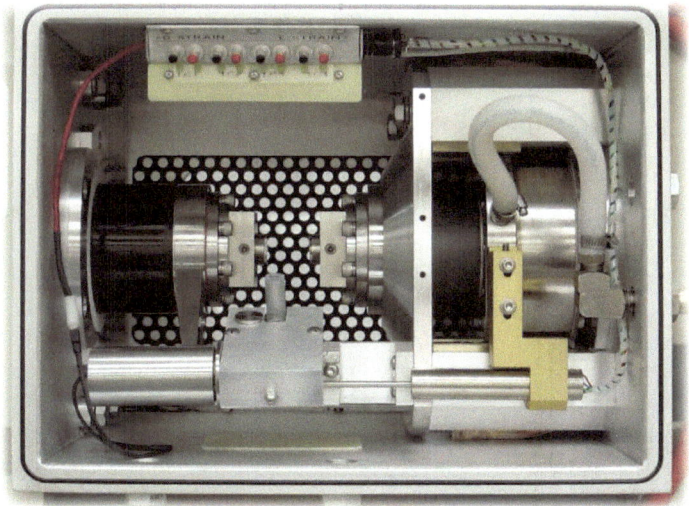

Fig. 2.7 Hydrawedge module working chamber

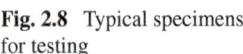
Fig. 2.8 Typical specimens
for testing

programming of a rolling mill operation. The results of the experiment are processed on the standard software Origin.

Each module has its own working chamber, which is connected to a vacuum system, providing a discharge of at least $1 \cdot 10^{-4}$ mm Hg. Art. Tests can be carried out at a lower vacuum, in a shielding gas or air.

The heating of the samples is carried out by direct transmission of an electric current, at a power of a welding transformer of 75 kVA; this provides a maximum heating rate of up to 12,000 °C/s. The temperature, heating and cooling rates are controlled by a thermocouple welded to the sample on the installation attached to the complex. Simultaneous recording of temperature at four points of the working part of the sample is possible, one of the thermocouples being the control one. The temperature control system provides its oscillation when heated at a rate of 1000 °C/s no more than 5–6 °C and maintenance at a given level with an error of not more than ± 1 °C.

During the experiments, samples are cooled in different ways:

- Heat sink in water-cooled copper or steel grippers;
- Blowing air or inert gas;
- water flow outside, inside the samples, or both inside and out.

To ensure the tests, power supply, hydraulic oil, compressed air and distilled cooled water are supplied to the main unit and modules.

The maximum cooling rate achieved when testing the complex on samples with a thickness of 6 mm is 4500 °C/s. When the samples are cooled by air or water, the vacuum system is disconnected from the working chamber.

Tests can be carried out at varying temperatures, from room temperature to melting point. Moreover, special quartz tubes are used for studying the liquid–solid state.

To record forces and deformations, strain gauges and displacement sensors of the movable traverse are used, respectively. The tensile and compression test module is equipped with highly sensitive meters for longitudinal and transverse deformation, as well as with a dilatometer; all of them allow performing highly accurate measurements of displacements and determining the temperatures of phase transformations in a given range of heating and cooling rates after plastic deformation or without it.

Maximum deformation rates up to $200 \, s^{-1}$ are achieved when tested on specimens with dimensions of Ø 10 mm × 15 mm using a shock test module for which the speed of the working crosshead is 2.5 m/s. Thus, the deformation rates on the Gleeble-3800 cover the entire range of high-speed deformation modes of the most modern hot rolling mills.

In this work, the plastic deformation of high-strength low-alloy steels was performed on the Pocket Jaw "tension–compression" module (Fig. 2.9) and on the MAXStrain module for multi-axis deformation (Fig. 2.10). The Pocket Jaw module realizes a uniaxial tension/compression strain and is characterized by the highest heating and cooling rates of the samples. The MAXStrain module implements multi-axis deformation. The difference of this module is the ability to alternate deformation in two mutually perpendicular directions by rotating the specimen 90° relative to its longitudinal axis (clockwise and counterclockwise).

In the first case, the heating was carried out at a rate of 100 °C/s to 1200 °C, the samples were cooled to T = 800, 850, 900, 950 °C, then deformed repeatedly at a constant temperature (800, 850, 900, 950 °C), then cooled naturally to room temperature. In the second case, the sample was also heated to 1200 °C, after which repeated deformation was performed at a decreasing temperature.

Samples for the plastometric studies on compression were billets of diameters 10 mm and height 12 mm, and for tensile testing—10 mm in diameter and 86 mm in length, made of steel not subjected to deformation processes (Fig. 2.11) [12]. The investigations were carried out over a wide range of temperature changes, deformation rate and deformation rate (T = 20–950 °C, ε = 0.1–0.9, u = 0.1–100 s^{-1}).

Processing of experimental data in the Gleeble-3800 simulator is performed using a special programming language Gleeble System Language. The values of the plasticity limit Λp of the steels studied were calculated using the Kolmogorov method

Fig. 2.9 The pocket jaw module

Fig. 2.10 MAXStrain module

Fig. 2.11 Samples for
plastometric studies on the
Gleeble-3800 complex:
a Compression method;
b Stretching method

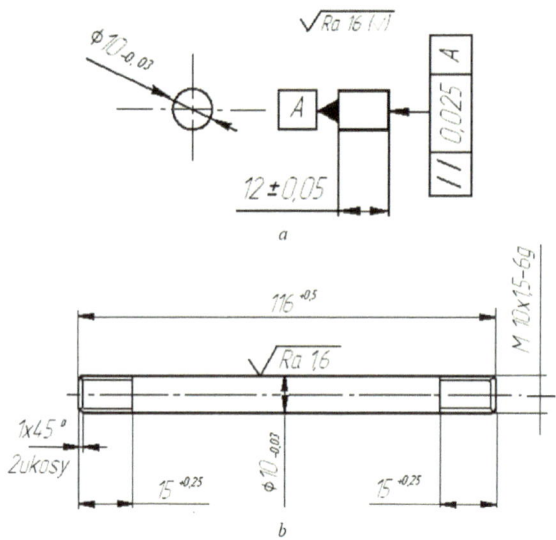

[13], based on the relative elongation and narrowing of the samples determined in
the tensile test [14, 15].

A standard Microsoft Office Excel program was used to process the final results
of the study. The program allowed a comprehensive analysis of the experimental
data obtained, which made it possible to reduce the number of samples to two for
one point, with a maximum error of 2–3%.

The change in the physicomechanical properties of the sample in the processes
of thermoplastic processing is a consequence of a significant restructuring of the
micro- and mesostructure of the material. It is impossible to describe such processes

without studying and creating appropriate models that explicitly take into account the physical origins of the evolution of the microstructure of the material under large deformations. Considerable attention is paid in physical theories to the modification of the hardening laws in connection with new experimental data obtained with the use of high-resolution equipment (in particular electron microscopes) and obtaining experimental hardening curves for large plastic deformations by the method of plastometry.

When tested on a plastometer, samples measuring $d \times h = 10$ mm $\times$ 12 mm were placed in a chamber inside which air was evacuated and a vacuum was created to prevent oxidation of the metal. The control of the plastometer was carried out by special computer programs in terms of temperature, speed, and degree of deformation. At certain intervals during the loading process, the yield stress and the logarithmic deformation were recorded. In Table. 2.1 shows the thermomechanical parameters of the deformed samples [16].

Figure 2.12 presents diagrams of sample compression at different temperatures, showing the dependence of the yield stress on the logarithm of the compression deformation at different stages of shaping. With an increase in the degree of deformation, σs only grows at a temperature of 850 °C and maintains the index at 900 °C and above. This is due to the flow—$\alpha - \gamma$ transformation (Fig. 2.12, curves 3, 2).

The resistance of deformation of 10HFTBch (Standard of Ukraine) steel with an increase in the deformation temperature from 800 to 850 °C does not increase significantly and remains almost unchanged only at high strain rates. This is due to the effect of two opposite factors: temperature, which reduces the resistance to deformation and structural resistance, which increases the resistance to deformation, as a result of an increase in the proportion of austenite in steel (Fig. 2.13a, b). As might be expected, with increasing deformation rate, the deformation resistance increases, especially with an increase in the proportion of the austenite component (Fig. 2.14a–c).

To develop the modes of plastic deformation ensuring the maximum possible grinding of the structure of low-carbon, low-alloy steels, detailed information is needed on the effect of austenite deformation regimes on the formation of structural elements in steel. To determine the critical degrees of deformation that ensure the formation of recrystallized austenite grain, studies were made of the influence of the degree and temperature of deformation on the structure of low-alloyed low-carbon steel. The deformation temperature was varied from 770 °C to 950 °C, the deformation rate was ln $\varepsilon = 1.2$, and the deformation rate was 100 s^{-1}. Microstructures and distribution diagrams for the grain score are shown in Figs. 2.15 and 2.16.

Table 2.1 Parameters of deformation, not less than

Samples	1	2	3	4	5
Temperature (°C)	770	800	850	900	950
Deformation rate (s^{-1})			100		
Degree of deformation, ln ε			0, 01, …1, 2		

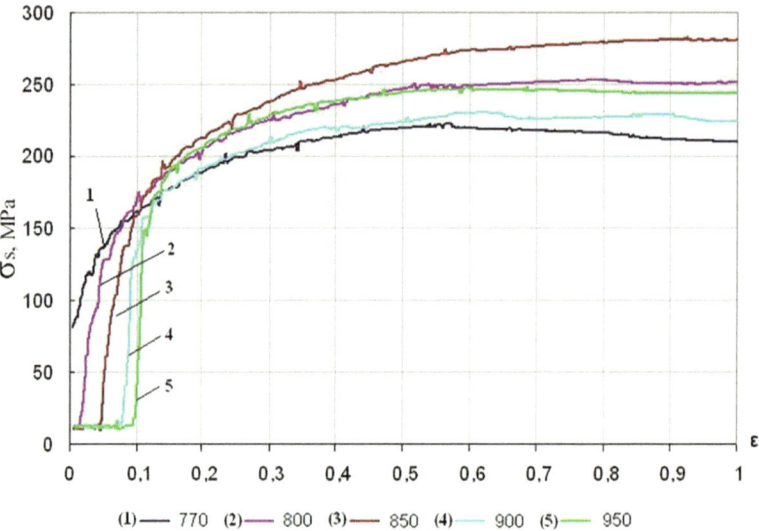

Fig. 2.12 Dependence of stresses on the initial stage of deformation

After deformation at a temperature of 770 °C, a ferrite-pearlite structure with an average grain size of 11 is formed in the steel, the maximum 14 points of grain is 2.42%, in the structure 28.34% are structural elements with a grain score of 10 (Fig. 2.17a). In steel, 81.5% of ferrite and 18.5% of perlite are formed. This is probably due to the fact that the deformation passes in a dual-phase region with the formation of a significant fraction of ferrite (Fig. 2.17b).

After deformation at a temperature of 850 °C, a structure with an average grain size of 12 is formed. In the structure, 33.35% of grains with a size of 11 points are observed, 31.09%—with a size of 12 points, the maximum size of 14 points of grain is 5.63% (Fig. 2.18a). 62.04% of ferrite and 37.95% of perlite are formed in the structure (Fig. 2.18b). This indicates the grinding of the α-phase—the formation of a subgrain structure inside the grain.

A further increase in temperature to 950 °C leads to an insignificant increase in the average size of the structural element after deformation $\ln \varepsilon = 1.2$ (Fig. 2.19a), in a structure of 24.52% is a grain score of 11, 20.16%—a grain score of 12. The maximum 14 points of grain size is 10.50%. 58.46% of ferrite and 41.53% of perlite are formed in the structure (Fig. 2.19b).

When considering the histograms of the distribution of structural elements in terms of grain size (see Figs. 2.17, 2.18 and 2.19a), one can note that the distribution pattern is approximately the same at different deformation temperatures. When analyzing the histograms of the distribution over the phases (Figs. 2.17, 2.18 and 2.19b), the differences are seen: at a deformation temperature of 770 °C, the peak of the distribution accounts for the fraction of ferrite and is more than 80%, at higher

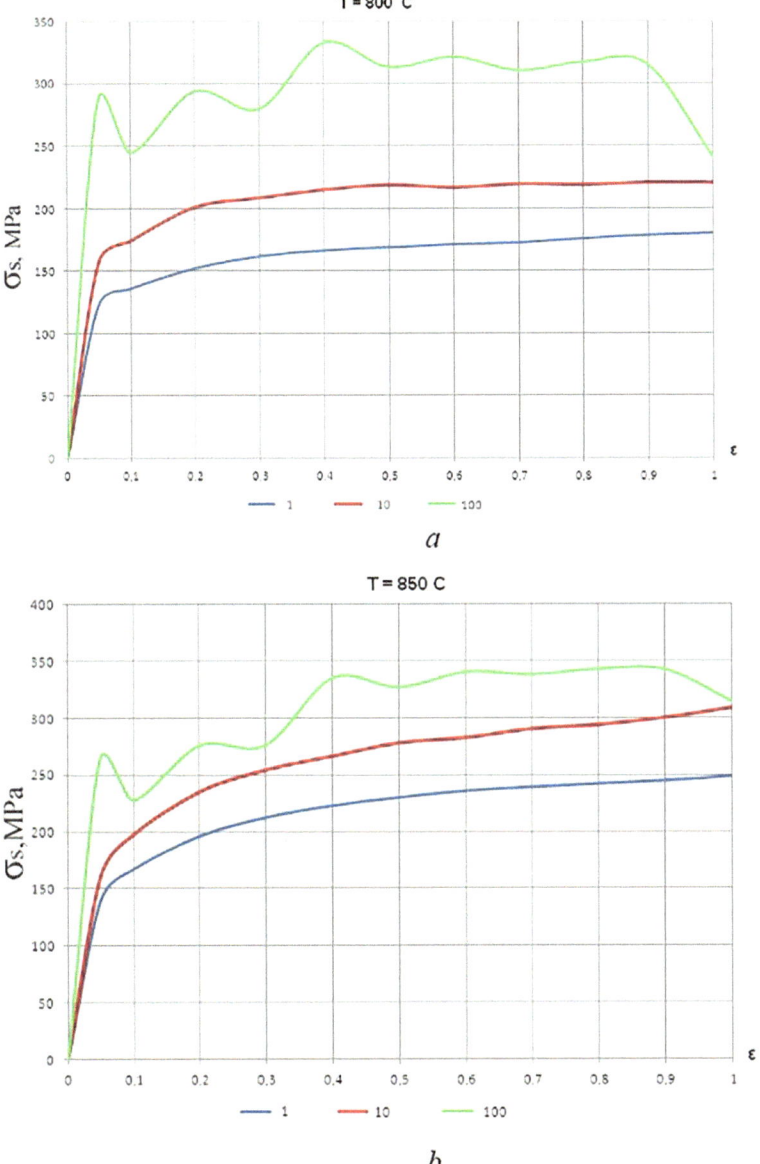

Fig. 2.13 Dependence of the deformation resistance on the degree of deformation at different temperatures: **a** 800 °C; **b** 850 °C

Fig. 2.14 Dependence of
the deformation resistance
on the deformation rate:
a 1 s^{-1}; **b** 10 s^{-1}, **v** 100 s^{-1}

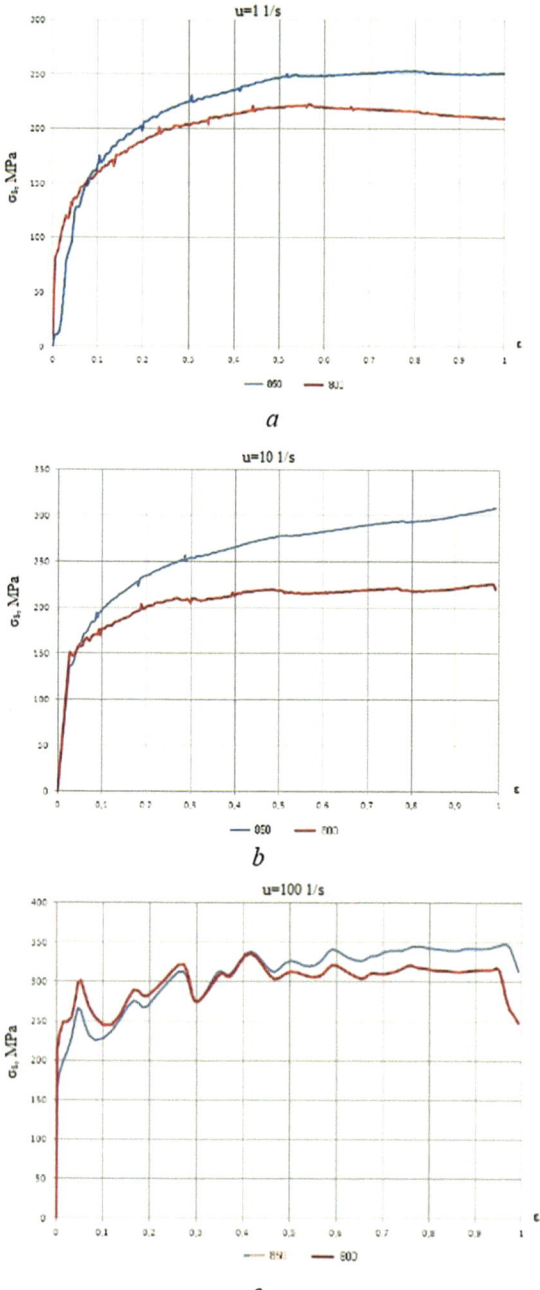

Fig. 2.14 Dependence of the deformation resistance on the deformation rate: **a** 1 s^{-1}; **b** 10 s^{-1}, **v** 100 s^{-1}

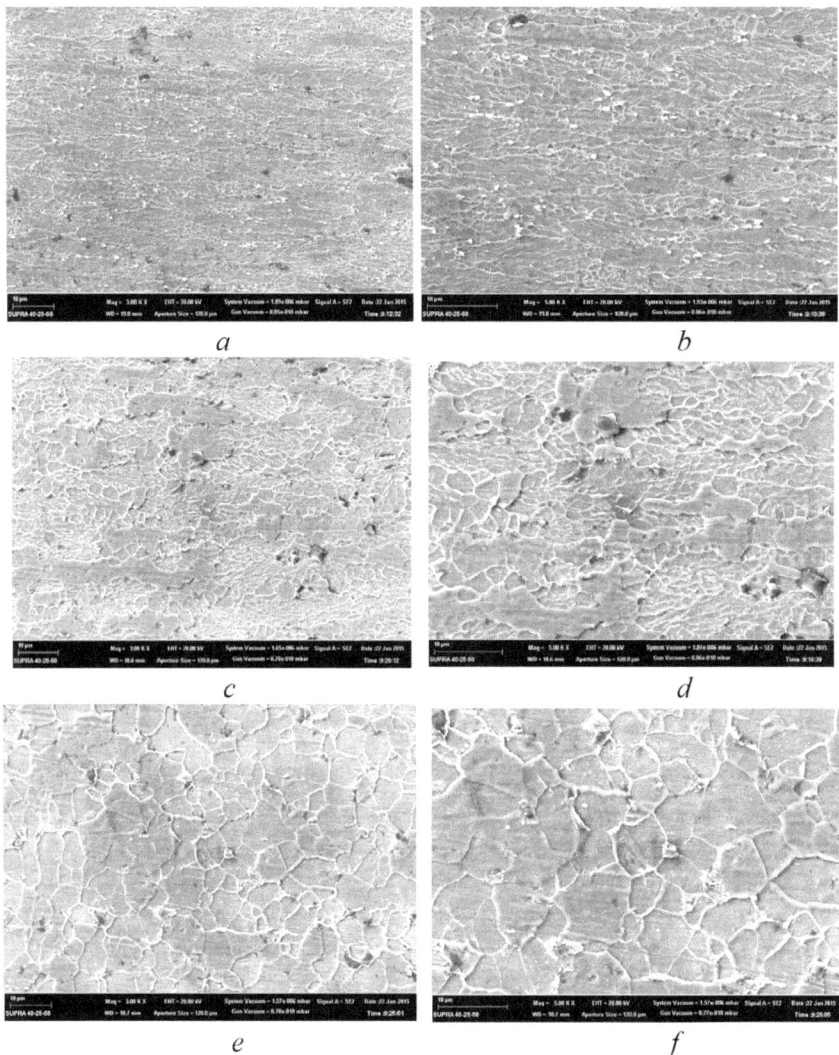

Fig. 2.15 Microstructures of low-alloy steel 10HFTBch (Standard of Ukraine) after deformation: **a, b** 770 °C; **c, d** 800 °C; **e, f** 850 °C; **a, c, e** × 3000; **b, d, f** × 5000; with a degree of ln $\varepsilon = 1, 2$ and a deformation rate of 100 s^{-1}

temperatures the fraction of ferrite decreases, and lies within the boundaries of 58–62%, which indicates a change in the ratio of structural constituents of ferrite (with predominantly large-angle misorientations between grains).

Microhardness was measured on the samples under study. The dependences of the change in the microhardness and the average size of the structural element on the deformation temperature are established. On a sample deformed with a deformation rate ln $\varepsilon = 1, 2$ at a temperature of 770 °C, the microhardness value is the lowest

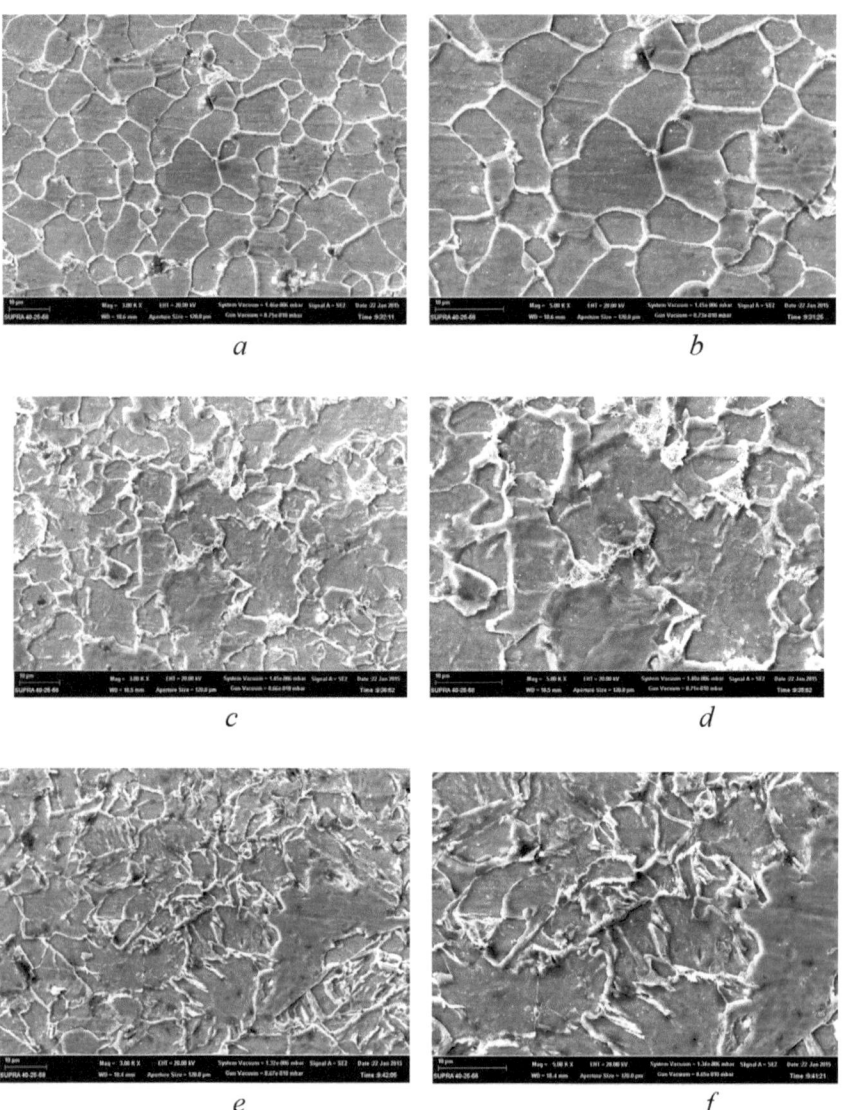

Fig. 2.16 Microstructures of low-alloy steel 10HFTBch (Standard of Ukraine) after deformation: **a, b** 900 °C; **c, d** 950 °C; **e, f** 1100 °C; **a, c, e** × 3000; **b, d, f** × 5000; with degree of ln $\varepsilon = 1, 2$ and deformation rate of 100 s^{-1}

and amounts to 260 HV, an increase in temperature to 850 °C leads to an increase in hardness up to 320 HV, a further increase in temperature does not affect the value of microhardness.

The lowest value of microhardness was obtained on samples deformed at a temperature of 770 °C, although in these samples the most fine-grained structure is observed.

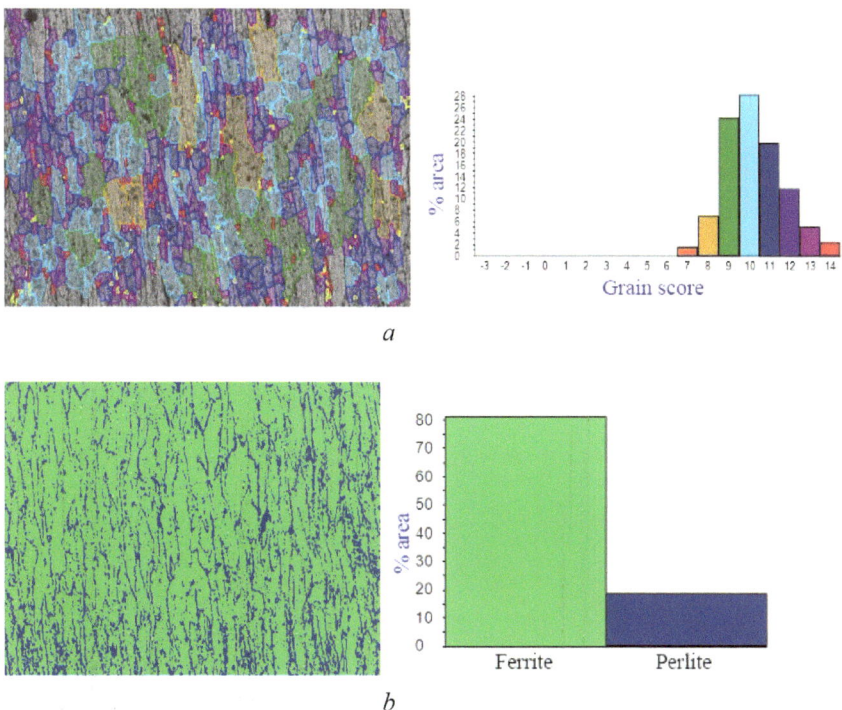

Fig. 2.17 Results of the study of the microstructure of steel: **a** the distribution of the grain score; **b** phase distribution; after deformation $\ln \varepsilon = 1, 2$ at a temperature of 770 °C and a strain rate of $100\ s^{-1}$, × 250

Perhaps this is due to the fact that plastic deformation passes in a two-phase region and during deformation a large amount of ferrite is formed, which differs in low strength (hardness).

The data obtained are in good agreement with modern concepts of the mechanisms of structure formation as a function of the temperature of deformation of steel. According to these ideas [17], at a deformation temperature of 770 °C, the predominant mechanism of fragmentation of the structure is fragmentation, which consists in breaking the austenitic and also ferritic grains of a single initial orientation into disoriented subgrains (fragments) by small-angle dislocation boundaries of deformation origin. At higher temperatures, two competing mechanisms are realized: fragmentation and initial processes of dynamic austenite recrystallization. Due to the formation of ferrite at a temperature of 770 °C, the minimum values of hardness are observed, the maximum hardness values in combination with the small size of the structural element are observed after deformation of 850 °C with a deformation degree of 1.2.

The effect of the degree of deformation on the structure formation of 10HFTBch steel was evaluated by hot rolling of samples (models). The results obtained from

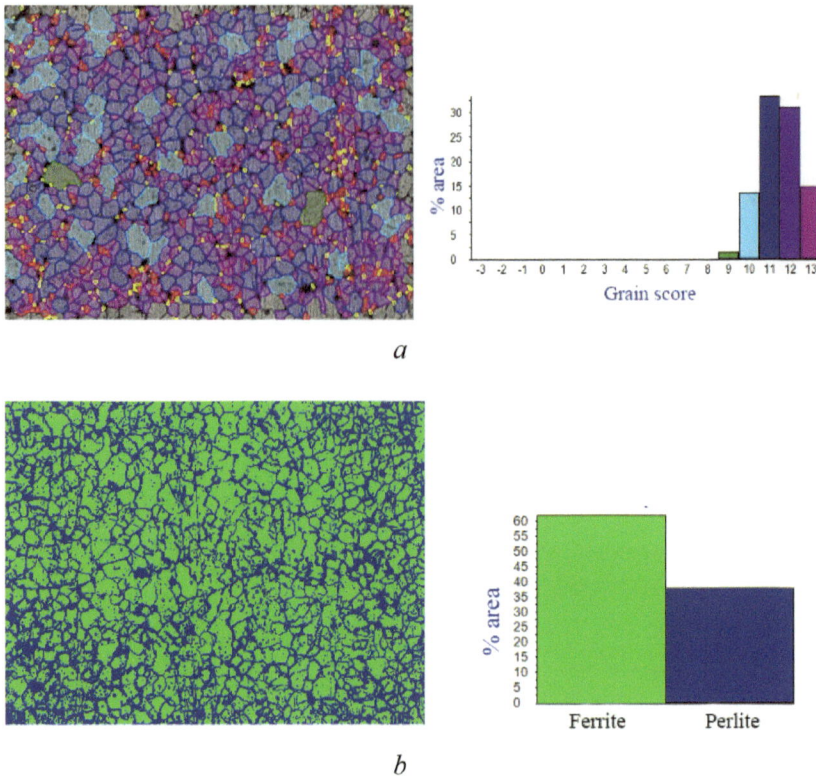

a

b

Fig. 2.18 Results of the study of the microstructure of steel: **a** the distribution of the grain score; **b** phase distribution; after deformation ln ε = 1, 2 at a temperature of 850 °C and a strain rate of 100 s^{-1}, × 250

the experiment carried out in Table 2.2 show that all samples rolled with different degrees of deformation had a high grain size.

The normalization carried out at 900 °C did not substantially eliminate the heterogeneity. An increase in the degree of deformation contributes to an increase in grain size from 4–6 points to 3–8 points, and an increase in the strain rate from 10 to 100 m/s did not lead to a noticeable change in grain size. Moreover, in the zone of intense deformation, a certain increase in grains was observed from 7 to 6 and even to 5 points. This is due to a more intense heating of the central section of the rolled section. The maximum heating of the third zone of the rolling section, measured with a thermocouple during hot deformation, increased by 25–30 °C [16].

The dependence between the grain size, temperature, and degree of deformation is shown in Fig. 2.20. At small degrees of deformation (usually, not more than 15%), the grain size does not depend on the degree of deformation. This section of the diagram is called the recrystallization threshold. With increasing temperature, the value of the recrystallization threshold decreases. At a certain critical degree of deformation

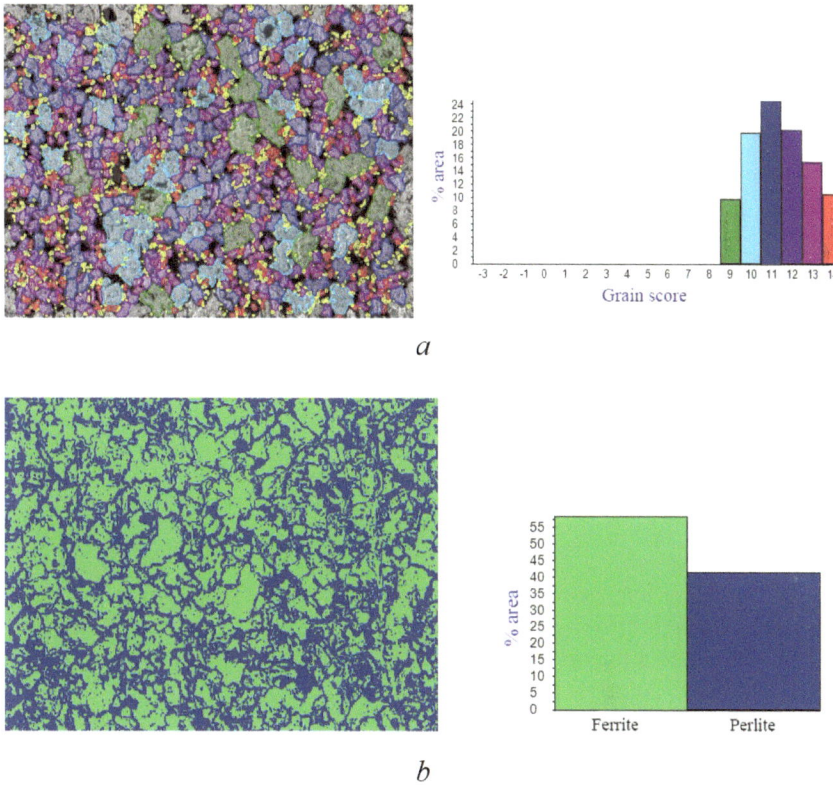

Fig. 2.19 Results of the study of the microstructure of steel: **a** the distribution of the grain score; **b** phase distribution; after deformation $\ln \varepsilon = 1, 2$ at a temperature of 950 °C and a strain rate of 100 s^{-1}, × 250

Table 2.2 Dependence of grain size on the degree and temperature of deformation of 10HFTBch (Standard of Ukraine) steel samples

T (°C)	Degree of deformation of samples[a]														
	$\varepsilon_1 = 0.85$			$\varepsilon_2 = 0.75$			$\varepsilon_3 = 0.65$			$\varepsilon_4 = 0.55$			$\varepsilon_5 = 0.45$		
	Grain score by sample zones, not less than														
	I	II	III	I	II	III	I	II	III	I	II	III	I	II	III
770	3–4	4–5	5	3–4	4–5	5	3–4	5	5	5	5	6	4	4–5	5
800	3	3–4	4–5	3–4	4–5	5	4	5	5	5	5	6	4	4	5
850	7	7	7–8	6–7	7–8	7	6–7	7	8	7	8	8	6–7	7	7
900	4	5–6	6	5–6	6	7	5–6	6	7	5	7	7	5	6	6
950	4	5	6	5	5	6	5	5	6	5	6	7	5	6	6

[a] The area of the original sample was $F = 20 \times 60 = 1200 \text{ mm}^2$. The samples were thermocoupled. After hot rolling, the samples were cooled in air

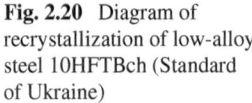

Fig. 2.20 Diagram of recrystallization of low-alloy steel 10HFTBch (Standard of Ukraine)

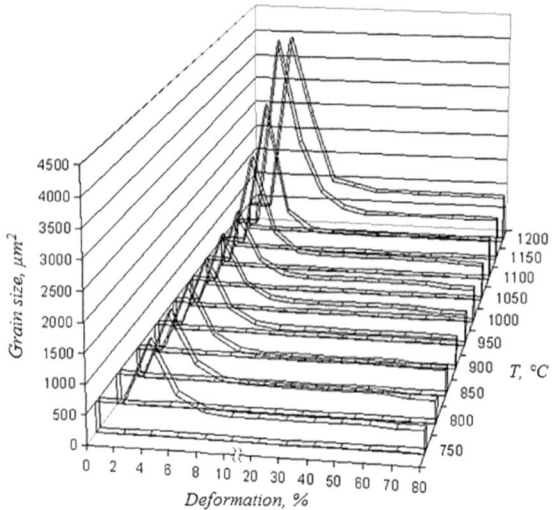

typical for a given temperature, the grain size will increase sharply, then with further growth of the deformation again decreases. This leads to an important conclusion: in order to obtain a fine-grained structure of steel, it is necessary to avoid critical degrees of deformation [16].

The form of the curve connecting the grain size with the degree of deformation is the result of the following circumstances. At small deformations (corresponding to the recrystallization threshold), the latter flows inside the grains, so that the boundaries between the grains are not violated. Therefore, direct contact of neighboring crystallites and mutual rearrangement of their lattice to form a common new grain is difficult. Under these conditions, the grain size does not depend on the degree of deformation. When the degree of deformation is equal to the critical one, the intercrystalline deformation begins, but the number of fractures of grains—the centers of crystallization is still small. In connection with this, the number of newly formed grains is also small, but their magnitude is significant (the critical value is correct) [18]. The propensity for grain growth is due to a decrease in the level of potential energy that occurs due to a reduction in the grain boundaries and their distortion. Later, with increasing deformation, the number of crystallization centers continues to increase, and the size of the grains due to mutual interference decreases with their growth.

In Fig. 2.21 it is shown that at high degrees of deformation the grain size increases with increasing temperature. This is explained by the fact that at high temperatures alignment of the orientation of all the crystallites occurs, due to which the small grains merge into larger ones. A sharp increase in grain size, especially after high degrees of deformation and at high temperatures, can be explained by the onset of the secondary recrystallization process. The recrystallization diagrams, taking into account the duration of the hot treatment, until the completion of the primary

Fig. 2.21 Diagram of recrystallization of low-alloy steel 10HFTBch (Standard of Ukraine) at high degrees of deformation

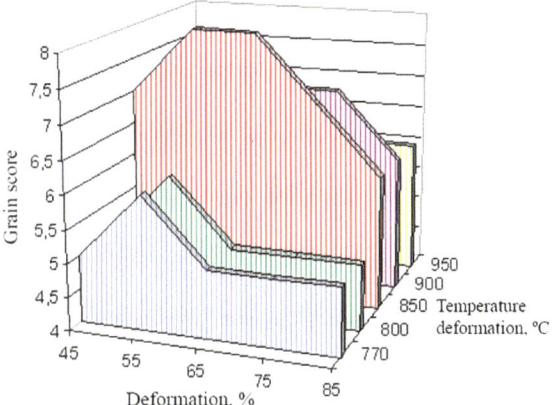

recrystallization, were established by H. Burgers and his co-workers [19]. The thermodynamic stimulus of secondary recrystallization is the desire of the system to reduce the total grain-boundary energy. One of the reasons for the strong inhibition of the growth of most grains is the dispersion particles of the second phase at their boundaries.

The results of the conducted studies show that the unevenness of deformation during hot rolling can substantially influence the process of structure formation at the moment of molding [19]. And the heterogeneity can both increase and decrease under the influence of temperature development in the focus of intense deformation. When developing the technological process of hot rolling low-alloy steels for the manufacture of car wheels, it is necessary to take into account the unevenness of the deformations along the section, and, accordingly, the increase in the heating of the central part of the slabs. In this case, a number of authors proposed the methods and derived the corresponding formulas, one of which (2.9) can be used to calculate the heating temperature of the central part of the slab [17]:

$$t = \theta \cdot (t_H - t_0) + t_0 \tag{2.9}$$

$$\theta k = 1 + \frac{\alpha \cdot 4 \cdot \sigma_{SH} \cdot L_H \left[ln \frac{H_H}{k} + \frac{1}{9} \left(\frac{L_k}{H_k} - \frac{L_H}{H_H} \right) \right]}{-\rho \cdot H_H \cdot (t_H - t_0)} \tag{2.10}$$

where

θ	dimensionless temperature;
t_H and t_0	temperature of initial deformation and environment, respectively;
α	coefficients of "heat output", $\alpha = 0.85 \div 0.95$.
σ_{SH}	the true resistance to deformation, contributing to the rate of deformation of nature $\acute{\varepsilon}_H$ and $\acute{\varepsilon}_0$;
H_H and L_H	height and length of the initial workpiece;

H_κ and L_κ height and length of the final rolled product;
C specific heat of steel;
P density of workpiece metal.

To determine σ_{SH}, in practice, expression (2.11) is used, which takes into account the strain rate of the workpiece

$$\sigma_{SH} = \sigma_0 + ln\frac{\dot{\varepsilon}_H}{\dot{\varepsilon}_0} \qquad (2.11)$$

where σ_{SH}—the true deformation resistance corresponding to the natural velocity $\dot{\varepsilon}_H$ and the sample $\dot{\varepsilon}_0$.

It should be noted that comparing the experimental results with the calculated results showed satisfactory convergence, which did not exceed 2.0%. The maximum heating of the central part of the samples under study, measured with thermocouples, at the time of hot deformation was 952 °C at the initial sample temperature of 900 °C.

Thus, it has been established that with an appropriate choice of thermoplastic deformation modes for steel of the same composition, higher strength or plastic properties can be obtained, which makes it possible to control the production of a given set of properties, and in the future, from a unified chemical composition, various strength categories, or sheet rolling with various visco-plastic properties, depending on the operating conditions.

2.3 Approximation of the Experimental Data of the Yield Stress of a New Steel Grade Under Different Thermomechanical Parameters

Modeling by the finite element method has become widely used as a mathematical tool for studying the stress–strain state in the processes of metal working with pressure. The finite element method is implemented in various software complexes, such as Deform-2D/3D, etc. For the correct statement of the problem, and therefore the adequacy of the results of calculations obtained, it is necessary to specify the rheological properties of the material in question in these software complexes. In this regard, the existence of a mathematical model that describes the rheological properties of the material for various deformation conditions is relevant from a practical and scientific point of view.

The conditions of deformation have a significant effect on the energy-force parameters of the process, the microstructure and the mechanical properties of the finished product. In the processes of metal working with pressure to assess the stress–strain state, energy-force parameters, an important characteristic is the deformation resistance.

To describe the change in yield stress, depending on the logarithmic deformation, temperature and strain rate, mathematical models are used, which are presented in [20–22].

In paper [22] 11 formulas are presented for the determination of yield stresses (Table 2.3).

Expressions (2.2)–(2.6) in a general form correctly reflect the influence of the thermomechanical parameters of the process on the yield strength σ_T of different steel grades. With an increase in the degree and rate of deformation, the stresses increase, decrease with increasing temperature.

Table 2.3 gives formulas for determining the yield stress σ_y. Expressions (2.1)–(2.9) confirm this dependence. The difference between yield stress and yield stress is that the yield stress determines the stress in the plastic zone at the beginning of loading, and the yield stress is determined during the loading process. The yield strength is a characteristic of the rheological properties of the material. In this regard, the above dependencies do not fully reflect the rheology of the material. They can not only grow with an increase in the degree of deformation, but also decrease or remain unchanged.

In this connection, expression 2.11, Table 2.3, the Hensel-Spittel formula [20]. Let's analyze it. Dependencies:

$$\alpha_1 \varepsilon^{\alpha_2};\ \exp(\alpha_4 \varepsilon);\ (1+\varepsilon)^{\alpha_5 T},$$

Table 2.3 Selected functions of yield stress for approximation

№	Formula
1	$\sigma_p = \alpha_1 + \alpha_2 \varepsilon + \alpha_3 u + \alpha_4 T$
2	$\sigma_p = \alpha_1 \varepsilon^{\alpha_2} u^{\alpha_3} \exp(-T\alpha_4)$
3	$\sigma_p = \alpha_1 + \alpha_2 \varepsilon + \alpha_3 u + \alpha_4 T + \alpha_5 \varepsilon u + \alpha_6 \varepsilon T + \alpha_7 Tu + \alpha_8 \varepsilon^2 + \alpha_9 u^2 + \alpha_{10} T^2$
4	$\sigma_p = \alpha_1 + \alpha_2 \varepsilon + \alpha_3 u + \alpha_4 T + \alpha_5 \varepsilon^2 + \alpha_6 u^2 + \alpha_7 T^2$
5	$\sigma_p = \alpha_1 + \alpha_2 \varepsilon + \alpha_3 u + \alpha_4 T + \alpha_5 \varepsilon u T^2$
6	$\sigma_p = \sqrt{3}\alpha_1 (\bar{\varepsilon} + \varepsilon_0)^{\alpha_2} (\bar{\varepsilon} + \bar{\varepsilon}_0)^{\alpha_3} \exp\left(\frac{\alpha_4}{T}\right)$
7	$\sigma_p = \alpha_1 \varepsilon^{\alpha_2} \exp(\alpha_3 \varepsilon) u^{\alpha_4} \exp(\alpha_5 T)$
8	$\sigma_p = \alpha_1 + \alpha_2 \varepsilon + \alpha_3 u + \alpha_4 T + \alpha_5 \varepsilon^2 + \alpha_6 u^2 + \alpha_7 T^2 + \alpha_8 \varepsilon^3$
9	$\sigma_p = \alpha_1 + \alpha_2 \varepsilon + \alpha_3 u + \alpha_4 T + \alpha_5 \varepsilon^2 + \alpha_6 u^2 + \alpha_7 T^2 + \alpha_8 \varepsilon^3 + \alpha_9 \varepsilon^4$
10	$\sigma_p = \alpha_1 \varepsilon^{(\alpha_2 + \alpha_3 T + \alpha_4 u)} \exp(\alpha_5 \varepsilon) u^{(\alpha_6 + \alpha_7 T)} \exp(\alpha_8 T)$
11	$\sigma_p = \alpha_1 \varepsilon^{\alpha_2} \exp\left(\frac{\alpha_3}{\varepsilon}\right) \exp(\alpha_4 \varepsilon)(1+\varepsilon)^{\alpha_5 T} u^{\alpha_6} u^{\alpha_7 T} T^{\alpha_8} \exp(\alpha_9 T)$

where
p yield strength;
ε degree of deformation;
u strain rate;
T deformation temperature;
$\alpha_1 \ldots \alpha_9$ constant coefficients

with an increase in the degree of deformation ε, the yield stress σp increases. Addiction:

$$\exp\left(\frac{\alpha_3}{\varepsilon}\right),$$

helps to reduce the yield stress.
Addiction:

$$u^{\alpha_6}; u^{\alpha_7 T},$$

with an increase in the strain rate, the yield stress increases.
Addiction

$$(1+\varepsilon)^{\alpha_5 T}; u^{\alpha_7 T}; T^{\alpha_8}; \exp(\alpha_9 T),$$

can increase or decrease the yield stress, from the ratios of thermomechanical parameters.

Thus, the Hensel-Spittel expression can describe the stress–strain curve, from thermomechanical parameters with different values change. The experimental data confirm this.

The graphs show that, in the case of small deformations ($\varepsilon = 0.2$–0.3), the yield strength increases strongly with increasing deformation. With average deformations ($\varepsilon > 0.3$), this growth of the yield point becomes less intense, and in a number of cases with a further increase in deformation, it decreases.

A preliminary analysis of mathematical models, carried out above, showed that the most acceptable formula for determining the yield stress at different thermomechanical parameters is:

$$\sigma = \alpha_1 \varepsilon^{\alpha_2} \exp\left(\frac{\alpha_3}{\varepsilon}\right) \exp(\alpha_4 \varepsilon)(1+\varepsilon)^{\alpha_5 T} u^{\alpha_6} u^{\alpha_7 T} T^{\alpha_8} \exp(\alpha_9 T). \tag{2.12}$$

To determine the coefficients in the formula (2.12) $\alpha_1 \ldots \alpha_9$, we can use the experimental data presented in Figs. 2.22, 2.23, 2.24, 2.25, 2.26, 2.27, 2.28 and 2.29.

A preliminary analysis of the experimental data shows that the mechanical characteristics of the new steel grade 10HFTBch (Standard of Ukraine) are largely determined by the thermomechanical parameters of plastic deformation: strain rate, deformation rate and temperature.

In Fig. 2.22 at T = 800 °C there are curves with different deformation rates: $u = 1\,s^{-1}$, $u = 10\,s^{-1}$, $u = 100\,s^{-1}$. With increasing strain rate, the yield stress increases. The nature of the change in the curves, with an increase in the relative reduction of ε, over the greater part of the graph is monotonically increasing. However, this increase, at ε from 0.2 to 1, is insignificant. At $u = 100\,s^{-1}$, fluctuations in the yield stress from $\varepsilon = 0.1$ are observed with gradual attenuation to $\varepsilon = 0.8$.

In Fig. 2.23 shows the curves at $T = 850$ °C for the same strain rates. There is a difference in the behavior of curves. As the strain rate increases, the yield stress

Fig. 2.22 Dependence of the yield stress on the logarithmic deformation at $T = 800\ °C$: $u = 1\ \mathrm{s^{-1}}$, $u = 10\ \mathrm{s^{-1}}$, $u = 100\ \mathrm{s^{-1}}$

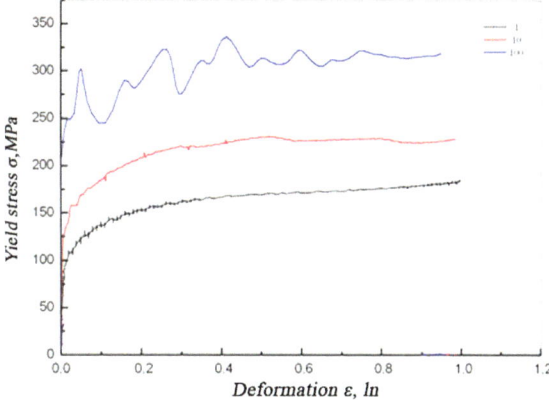

Fig. 2.23 Dependence of the yield stress on the logarithmic deformation at $T = 850\ °C$, $u = 1\ \mathrm{s^{-1}}$, $u = 10\ \mathrm{s^{-1}}$, $u = 100\ \mathrm{s^{-1}}$

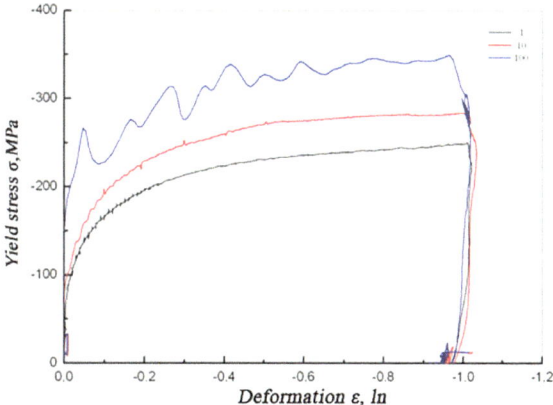

Fig. 2.24 Dependence of the yield stress on the logarithmic deformation at $T = 900\ °C$, $u = 1\ \mathrm{s^{-1}}$, $u = 10\ \mathrm{s^{-1}}$, $u = 100\ \mathrm{s^{-1}}$

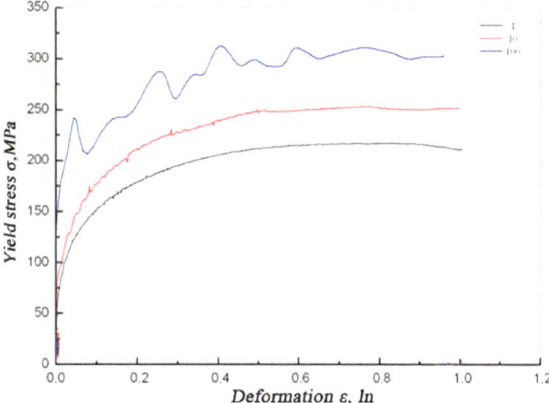

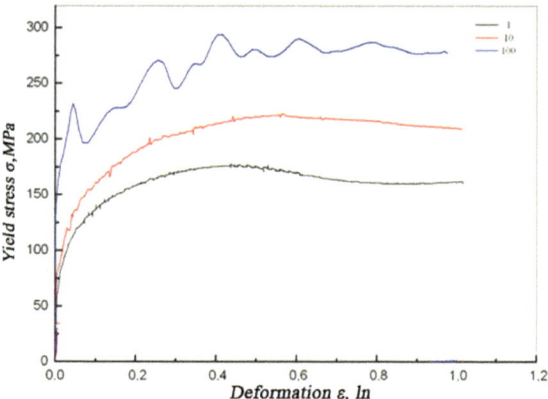

Fig. 2.25 Dependence of the yield stress on the logarithmic deformation at $T = 950\,°C, u = 1\,s^{-1}, u = 10\,s^{-1}, u = 100\,s^{-1}$

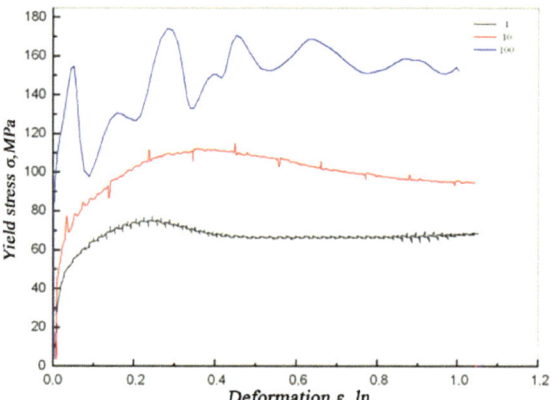

Fig. 2.26 Dependence of the yield stress on the logarithmic deformation at $T = 1200\,°C, u = 1\,s^{-1}, u = 10\,s^{-1}, u = 100\,s^{-1}$

increases, but the trajectory of the curve for $u = 100\,s^{-1}$ has a periodically changing character, with the curve aligned after the mark $\varepsilon = 0.7$.

In Figs. 2.24 and 2.25 are curves at $T = 900\,°C$ and $950\,°C$ at the same strain rates. At this temperature, unlike the previous graphs, the curves are located in accordance with generally accepted ideas about the effect of strain rate. As the strain rate increases, including $u = 100\,s^{-1}$, the yield stress increases. For $u = 100\,s^{-1}$, a periodically changing curve is observed.

In Fig. 2.26 data are presented at $T = 1200\,°C$, the curves are arranged in the appropriate sequence, with an increase in the strain rate, the yield stress increases. At a strain rate $u = 100\,s^{-1}$, significant deviations of the experimental curve are observed at $\varepsilon = 0.1$ to $\varepsilon = 0.5$.

In Fig. 2.27 shows the dependence of the yield stress for different temperatures, for $u = 1\,s^{-1}$. All the curves, except for the curve at $T = 800\,°C$, are arranged in an appropriate order, which is determined by the fact that with increasing temperature, the yield stress decreases. The curve at $T = 800\,°C$ should be located above all the others, however, in the main section the yield stresses turned out to be lower than on the curves with higher temperatures: $T = 850\,°C$, $T = 900\,°C$. For minor deformations, up to $\varepsilon = 0.01$, this increase is visible. When obtaining such conflicting data, repeated tests were carried out on the Gleeble 3800 plastometer. They confirmed the noted regularity. It can be assumed that such a change in mechanical characteristics at different temperatures with a low strain rate is a rheological feature for a given steel grade.

In Fig. 2.28 shows the distribution of the yield stress for different temperatures, with $u = 10\ s^{-1}$. The same feature was repeated in the characters of the curves, as in the case $u = 1\ s^{-1}$.

In Fig. 2.29 shows the dependence of the yield stress on the degree of deformation, at the same temperatures, the strain rate $u = 100\ s^{-1}$. There is a periodic change in the curves at $T = 800\,°C$, $T = 850\,°C$. The curve at $T = 800\,°C$ is somewhat higher than the curve at a temperature $T = 850\,°C$. In general, the trajectory of these curves for $u = 100\ s^{-1}$ behaves approximately identically.

A system of equations is created, the solution of which are the indicated coefficients.

Fig. 2.27 Dependence of the yield stress on the logarithmic deformation for $u = 1\ s^{-1}$, $T = 800\,°C$, $T = 850\,°C$, $T = 900\,°C$, $T = 950\,°C$, $T = 1200\,°C$

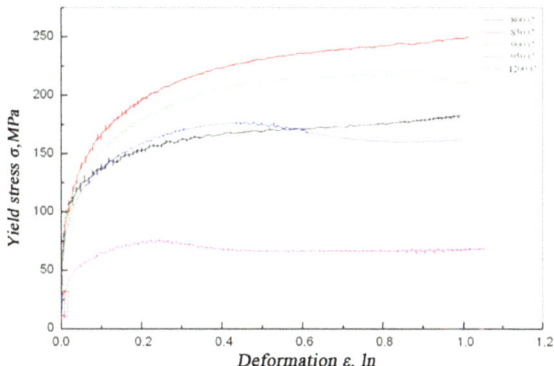

Fig. 2.28 Dependence of the yield stress on the logarithmic deformation at $u = 10\ s^{-1}$, $T = 800\,°C$, $T = 850\,°C$, $T = 900\,°C$, $T = 950\,°C$, $T = 1200\,°C$

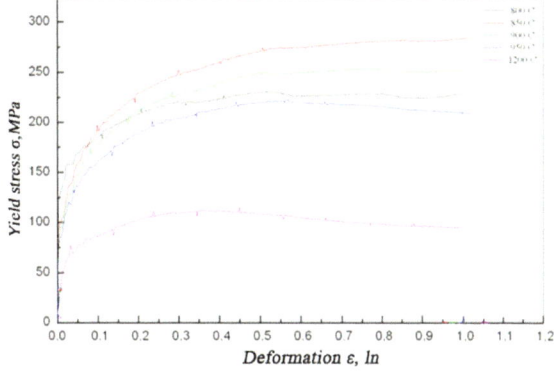

Fig. 2.29 Dependence of the yield stress on the logarithmic deformation for $u = 100 \text{ s}^{-1}$, $T = 800$ °C, $T = 850$ °C, $T = 900$ °C, $T = 950$ °C, $T = 1200$ °C

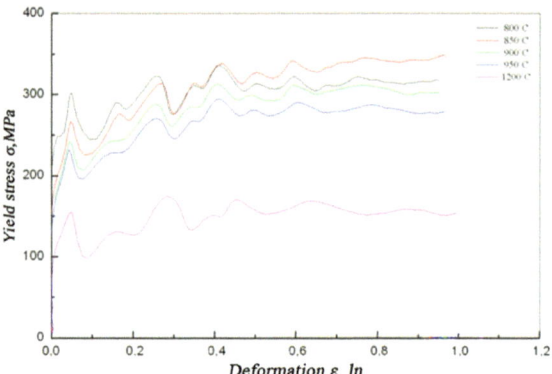

For example, at $T = 850$ °C, $\varepsilon = 0.4$, $u = 1$ s-1, $\sigma_y = 225$ MPa and at $T = 900$ °C, $\varepsilon = 0.6$, $u = 10 \text{ s}^{-1}$, $\sigma_y = 250$ MPa. We obtain a system of equations:

1)
$$225 = \alpha_1 0.4^{\alpha_2} \exp\left(\frac{\alpha_3}{0.4}\right)$$
$$\exp(\alpha_4 0.4)(1 + 0.4)^{\alpha_5 \cdot 850} 1^{\alpha_6} 1^{\alpha_7 \cdot 850} 850^{\alpha_8}$$
$$\exp(\alpha_9 850)$$

2)
$$250 = \alpha_1 0.6^{\alpha_2} \exp\left(\frac{\alpha_3}{0,6}\right)$$
$$\exp(\alpha_4 0.6)(1 + 0.6)^{\alpha_5 \cdot 900} 10^{\alpha_6} 10^{\alpha_7 \cdot 900} 900^{\alpha_8}$$
$$\exp(\alpha_9 900)$$

3) ....

n) ....

This system of equations, using the method of experiment planning, is calculated in such a way as to determine the coefficients $\alpha_1 \ldots \alpha_9$. The revised experimental data, in accordance with the Reology program, allowed us to determine the coefficients α_i (Table 2.4).

In Figs. 2.30 and 2.31. Comparative curves (approximating and experimental) are presented for different temperatures and strain rates.

The analysis of the graphs shows the comparability of the experimental and theoretical data, with the exception of the curve, at $T = 1200$ °C. At a strain rate $u = 100 \text{ s}^{-1}$, a significant deviation in the practical curve is observed, from $\varepsilon = 0.1$ to $\varepsilon = 0.5$.

Table 2.4 Coefficients obtained by approximation

α_1	α_2	α_3	α_4	α_5	α_6	α_7	α_8	α_9
0.000 082	0.524 152	− 0.000 163	1.113 630	− 0.003 363	−0.216 846	0.000 336	2.970 980	− 0.004 952

Fig. 2.30 Comparison of the dependencies of the yield stress on deformation $T = 850\,°C$

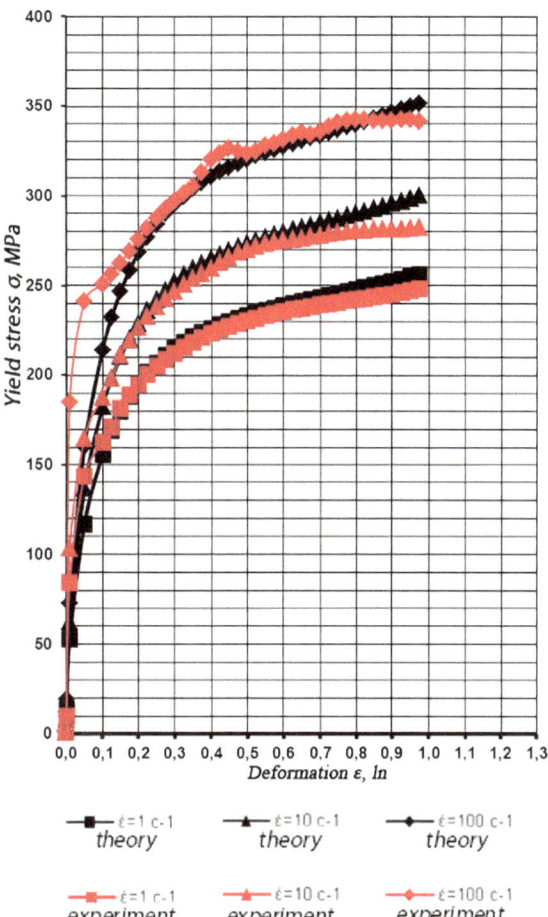

Thus, the peculiarity of this approximation is that with the help of the obtained formula it is possible to take into account the rheology of various steel grades. The dependence of the yield stress on deformation can be increasing, decreasing or not changing.

Such a formula can be used to calculate the energy-strength parameters and the yield stress of the metal at each point of the deformation center.

As a result of mathematical processing, an equation was obtained showing the dependence of the resistance of the deformation metal on the conditions of thermoplastic processing [23]:

$$\sigma_y = 0.000082\varepsilon^{0.524152}\exp\left(\frac{-0.000163}{\varepsilon}\right)$$
$$\exp(1.11363\varepsilon)(1+\varepsilon)^{-0.003363T}.$$

Fig. 2.31 Comparison of
the dependencies of the yield
stress on deformation $T =$
900 °C

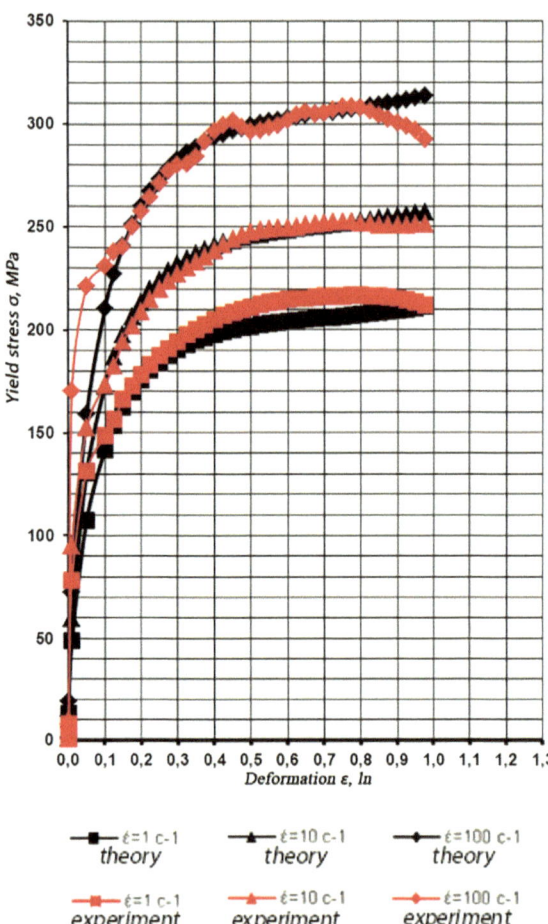

Fig. 2.31 Comparison of the dependencies of the yield stress on deformation $T = 900$ °C

$$\cdot\, u^{-0.216846}\, u^{0.000336T}\, T^{2.97098}\, \exp(-0.004952T). \tag{2.13}$$

To estimate the relationship between the experimental data on the graph and the Hensel-Spittel model, we calculated the correlation coefficient r, which makes it possible to estimate the tightness of the coupling of the variables, and also to determine what proportion of the changes are due to the influence of the main factor and other factors. The coefficient varies from -1 to $+1$. If $r = 0$, then the relationship between the signs is absent and speaks only of the absence of linear correlation dependence, but not in general about the absence of correlation, and even more so of statistical dependence. If $r = \pm 1$, then this means having a complete (functional) connection. The practical significance of the correlation coefficient is determined by its squared value, called the coefficient of determination.

The selective correlation coefficient is given by:

$$r = \frac{\sum_{i=1}^{n}(x_i - \overline{x})(y_i - \overline{y})}{\sqrt{\sum_{i=1}^{n}(x_i - \overline{x})^2 \cdot \sum_{i=1}^{n}(y_i - \overline{y})^2}} \qquad (2.14)$$

where

x_i, y_i	variants (observed values) of characteristics X and Y;
n	sample size;
$\overline{x}, \overline{y}$	sample mean;
$(x_i - \overline{x})(y_i - \overline{y})$	deviation of the individual variant from the total arithmetic mean of x and y.

To calculate the correlation coefficient, use the corresponding function in Excel.

It is established that the Hensel-Spittel model is characterized by the following correlation coefficients: 0.9713; 0.9661; 0.9613, which adequately reflects the existing relationship and can be used to model the rheological properties of 10HFTBch steel (Figs. 2.30, 2.31, 2.32, 2.33, 2.34 and 2.35).

To analyze the established regularities of the change in the resistance of the deformation metal, three-dimensional graphical dependences were constructed (Figs. 2.36, 2.37 and 2.38).

As the temperature rises, the amplitude of the thermal vibrations of the atoms increases and, accordingly, all strength characteristics, including the deformation resistance, decrease. Each of these curves has its own peculiarities, which in some narrow temperature intervals violate the monotonous character of the change in the resistance of deformation from temperature. Such a picture is observed if, with an increase in temperature, physical and chemical processes occur in the steel, which can lead to an increase in σs in the temperature range of these processes [19].

The temperature relationship is subject to the condition $T_1 < T_2 < T_3$, i.e. the upper curve corresponds to the minimum value of T. According to the graph (Fig. 2.36), it is seen that with increasing temperature, the degree of deformation at which softening begins begins to decrease.

On the basis of experiments, it was established that in the temperature range under consideration, the change in σ_s with variation of T is subject to an exponential dependence.

The ratio of strain rates obeys the condition $u_1 > u_2 > u_3$, and the degree of deformation at which the softening occurs is almost independent of the strain rate (see Fig. 2.37). Taking into account the effect of the degree and strain rate on σs is complicated by the fact that the temperature of the softening processes decreases with increasing degree of deformation, the heat output, and the temperature of the deformed body increase. In turn, an increase in the rate of deformation also contributes to an increase in temperature, reducing heat loss to the environment [23]. Thus, increasing the speed and degree of deformation directly increases the resistance to deformation, and indirectly, on the contrary, reduce it (see Fig. 2.38).

When hot processing, when the heat output is small, the strengthening effect of both factors prevails, and the deformation resistance increases with increasing degree and strain rate.

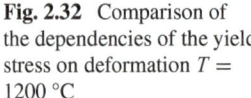

Fig. 2.32 Comparison of the dependencies of the yield stress on deformation $T = 1200\ °C$

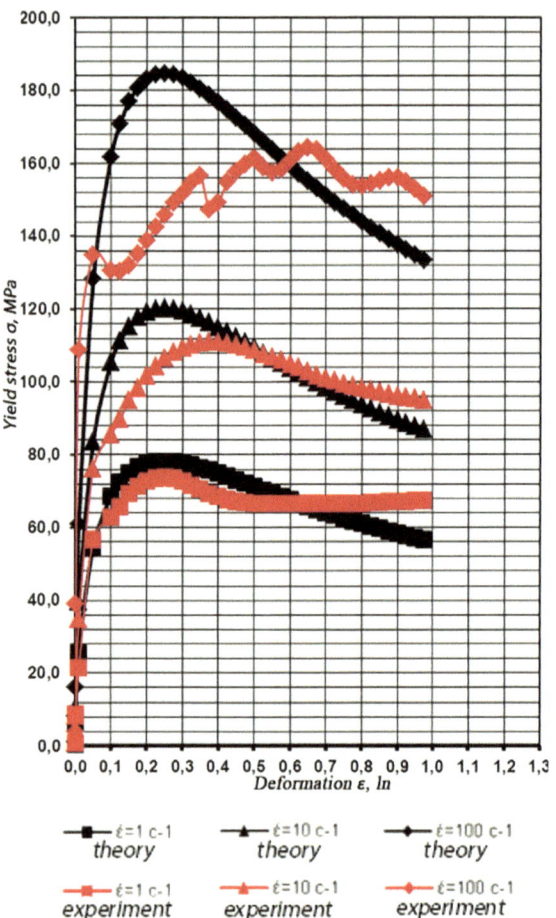

Almost all metals and alloys, the deformation resistance in the region of small deformation increases with increasing ε.

Thus, at the same rate of hardening and softening, the resistance to deformation does not depend on the degree of deformation. Sometimes the value of σs for large degrees of deformation may be less than the initial value for the same other deformation conditions. The degree of deformation at which softening occurs is almost independent of the rate of deformation. As the temperature increases, the rate of deformation decreases.

The Hensel-Spittel model is characterized by the following correlation coefficients: 0.9713, 0.9613, 0.9661, which adequately reflects the existing relationship and can be used to model the rheological properties of steel. The optimum temperature, degree and speed of hot deformation, providing the increased mechanical properties of the developed steel 10HFTBch are determined.

Fig. 2.33 Comparison of the dependencies of the yield stress on deformation $u = 1 \, s^{-1}$. *Source* Sheyko et al. [24]

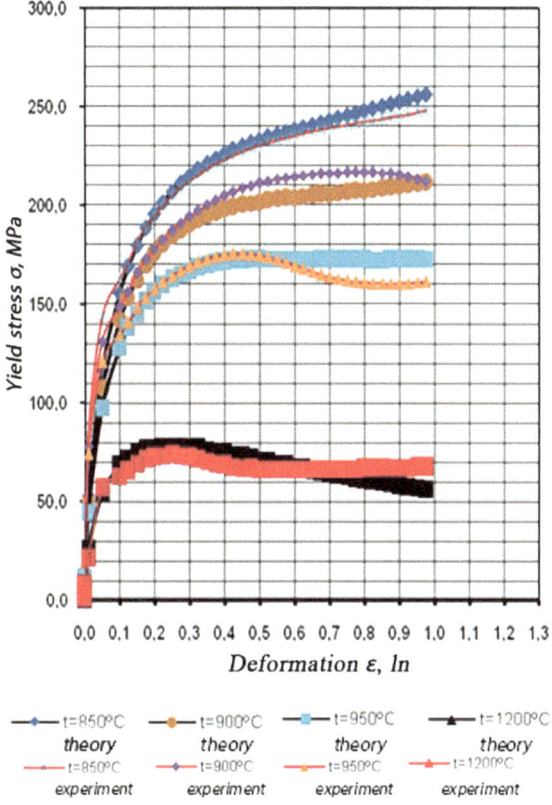

2.4 Effect of Hot Rolling on the Formation of the Structure and Properties of Low-Alloy Steel 10HFTBch

To assess the effect of hot rolling, under conditions of production of deformation of the structure and properties of hot-rolled plate, a blank was chosen with the parameters of an industrial slab: 135 mm thick, the width of a barrel of a laboratory rolling mill was 150 mm wide.

To reduce costs, the study of the features of hot slab deformation was carried out by modeling using the principle of similarity. The interaction between the rigid roll and the deformable material of the hot workpiece was simulated by contact surfaces that describe the contact conditions between the roll surfaces and the surface of the sheet. During the simulation, the contact conditions are constantly updated, reflecting the rotation of the rolls and deformation of the material, which makes it possible to model the sliding between the roll and the material of the workpiece being machined. The contact between the roll and the strip is modeled by friction along the Coulomb, the coefficient of friction was taken as 0.5. The initial grain size of the austenite, after heating in the furnace, was 200 μm [25, 26]. The time for interidentification pauses

Fig. 2.34 Comparison of the dependencies of the yield stress on deformation $u = 10\,s^{-1}$. *Source* Sheyko et al. [24]

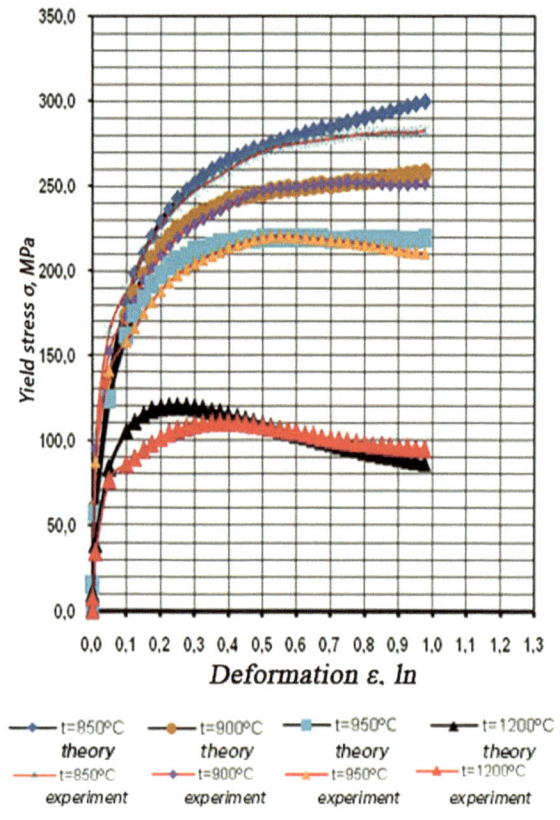

was maintained in the interval 1.6–15.0 s. The deformation resistance of low-alloy steel 10HFTBch is described by the obtained Eq. (2.13).

The grain size of austenite d_γ of low-alloyed steel, depending on the initial grain size d_0, holding time t and temperature T, is described by the equation [25, 26]:

$$d_\gamma = \left[d_0^3 + 5,47 \times 10^{20} t \exp\left(-\frac{460{,}000}{RT} \right) \right]^{\frac{1}{3}} \tag{2.15}$$

For hot rolling, the passage of metadynamic or static recrystallization is characteristic [27]. One of the methods for modeling recrystallization is the expression "Kolmogorov-Johnson-Mall-Avrami". For the static recrystallization of low-alloy steel, the recrystallized volume is determined by the expression [25, 26]:

$$X_{srex} = 1 - \exp\left[-0.693 \left(\frac{t}{t_{0,5}} \right)^{0,5} \right] \tag{2.16}$$

Time required to pass 50% static recrystallization of low-alloy steel [25, 26]:

Fig. 2.35 Comparison of the dependencies of the yield stress on deformation $u = 100 \text{ s}^{-1}$. *Source* Sheyko et al. [24]

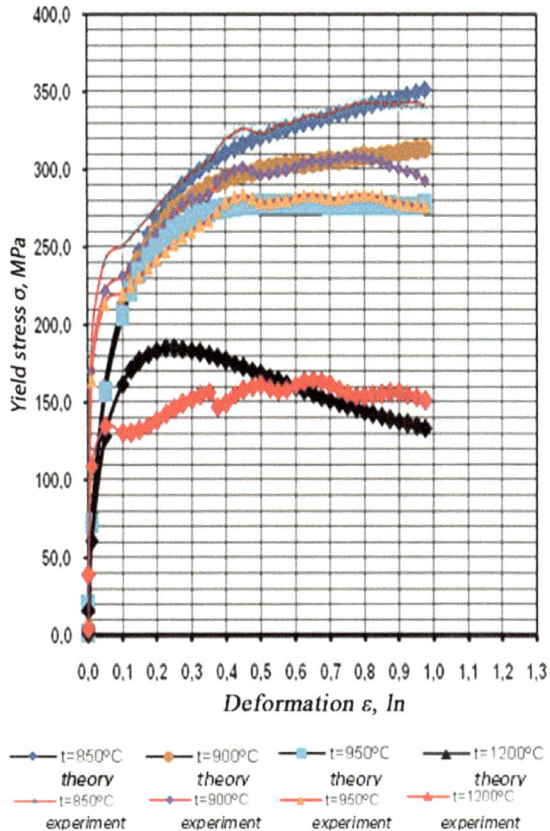

Fig. 2.36 Effect of temperature and degree of deformation on the character of the deformation resistance of 10HFTBch steel

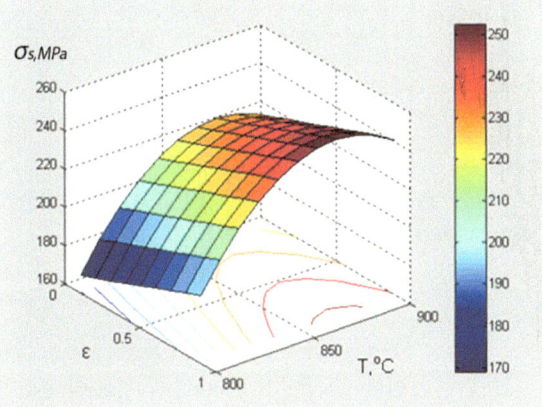

Fig. 2.37 Effect of temperature and strain rate on the character of the deformation resistance of 10HFTBch steel

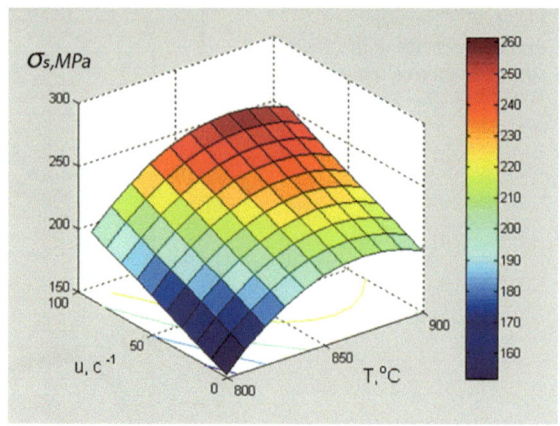

Fig. 2.38 Influence of strain rate and degree of deformation on the character of the deformation resistance of 10HFTBch steel

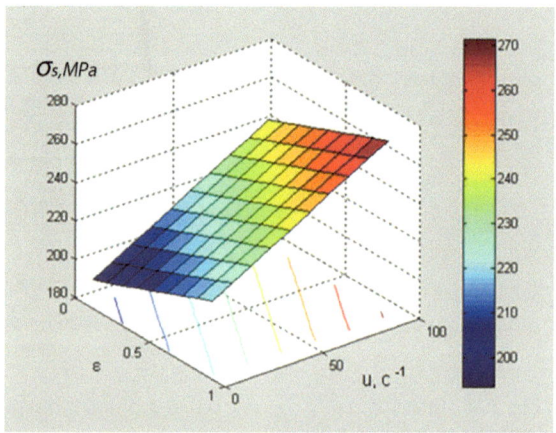

$$t_{0,5} = 4,92 \cdot 10^{-17} d^1 \varepsilon^{-2} \dot{\varepsilon} \exp\left(\frac{338,000}{RT}\right). \tag{2.17}$$

After the static recrystallization of low-alloyed steel, the grain size can be determined using the expression [25, 26]:

$$d_{rex} = 1200 d_0^{0.33} \Gamma^{-0.79} \exp\left(-\frac{88,000}{RT}\right). \tag{2.18}$$

In Fig. 2.39 shows the distribution of the intensity of deformations (Γ) and stresses (σ_i) during the rolling of steel 10HFTBch in the finishing group of the stands of the CBHRM 1680 mill.

By numerical simulation, the results obtained made it possible to establish that:

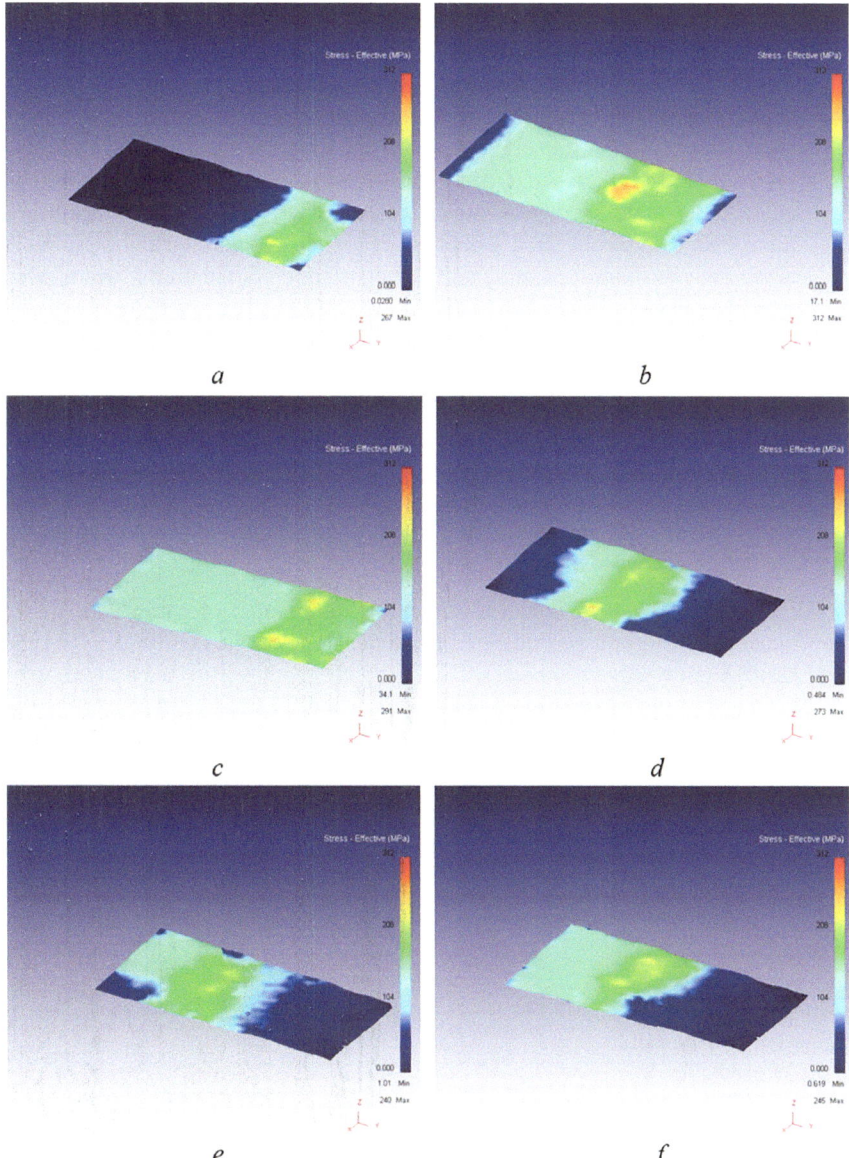

Fig. 2.39 The pattern of stress intensity distribution in the strip during rolling in the finishing group of the mill (the second stage of rolling in each stand): **a** stand no. 1; **b** no. 2; **c** no. 3; **d** no. 4; **e** no 5; **f** no. 6 [23]

(1) During rolling in the first stand of the proposed mill, the strain intensity (Γ) and stress (σ_i) at the initial moment of rolling are concentrated in the zones of metal gripping by the mill rolls. With increasing compression, the accent Γ and σ_i are transferred from the surface to the center and edges of the deformed workpiece;

(2) Deformations in the following stands of the finishing group of the mill, allow gradually to move the site to concentrate the intensity of deformation from the center to the middle part of the strip, and then to the contact zone of the roll with the rolled workpiece. Such a distribution of stress intensity and deformation along the cells leads to a more uniform distribution of the total Γ and σ_i along the deformation center;

(3) The most uniform distribution of the total Γ and σ_i in the height and length of the rolled strip was obtained during rolling with a single reduction in the first stand 38–45%, in the second stand 35–40%, in the third stand 33–38%, in the fourth stand 28–30%; in the fifth stand 22–25%; in the sixth stand 11–14%;

(4) During the rolling process in the first stand, the temperature in the hot metal-roll contact zones decreases. Rolling in subsequent stands allows, due to the release of heat of deformation and friction, to equalize the temperature along the deformation center.

Using the obtained picture of the intensity distribution of strain and stress, a dependence of the grain size change on the deformation parameters was obtained (Fig. 2.40).

As a result of the calculation of grain sizes, it was established that:

(1) After rolling in the first stand, the structure in the center of the strip is fine-grained and the grain size of the austenite is 70–85 µm, while in the surface zones of the strip the grain size of the austenite is more coarse (130–145 µm);

(2) Rolling in the subsequent stands of the finishing group of the mill, allows gradually to level out the dimensions of the austenite grain, which are equal to 55–65 µm throughout the section of the rolled strip.

Fig. 2.40 Change in the grain size of 10HFTBch steel during rolling in the finishing group of stands [23]

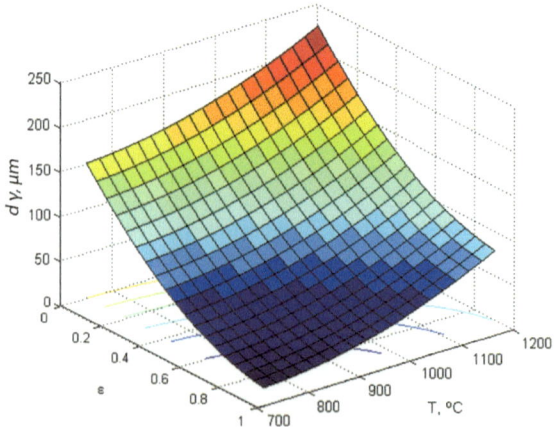

(3) The reduction in the rolling temperature to 850 °C promotes a certain refinement of the ferrite grain and a decrease in the size of the pearlite colonies. Thus, the average grain size of ferrite, samples rolled at 900 °C, was about 45–50 μm, and after rolling at 850 °C—about 30–36 μm. The ratio of structural components under the selected deformation conditions is: ferrite about 80%, perlite—about 20%.

An experimental verification of the simulation results and the study of the effect of the hot-strain scheme of 10HFTBch steel were carried out by rolling the samples (models) and studying their structure (Fig. 2.41). In the general case, it is possible to distinguish three zones with different degrees of deformability. In the first zone, the difference between the stresses is insignificant and may not meet the plasticity condition, although the stresses (levels) themselves are large. In the third zone, the restraining effect of frictional forces is the same as in the second, but the stress state scheme is different (there is a tensile stress), which reduces the plastic properties and worsens the deformability of the body.

The results of the experiment show that all rolled samples with different deformation temperatures had a different grain size over the rolled section (Figs. 2.42, 2.43, 2.44, 2.45, 2.46, 2.47, 2.48, 2.49, 2.50 and 2.51).

Fig. 2.41 Scheme of arrangement of fields of deformation: I—zone of inhibition; II—zone of hindered deformation; III—zone of intensive deformation

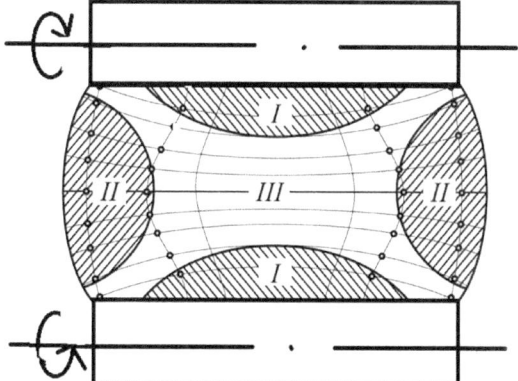

Fig. 2.42 Macro structure of the deformed sample at 770 °C

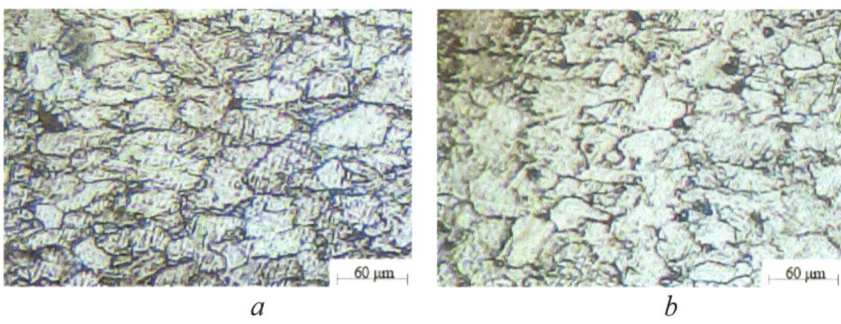

Fig. 2.43 Microstructure of deformed steel samples at a deformation temperature of 770 °C (×360): **a** contact zone with the tool, I; **b** zone of periphery, II

Fig. 2.44 Macro structure of the deformed sample at 800 °C

It is established that a decrease in the deformation temperature leads to a decrease in the dimensions of the recrystallized austenite grain, and consequently, the refinement of the ferritic grain. Also an important factor in preventing the growth of ferritic grains in the upper part of the ferrite region is the cooling of steel in rolls.

The recommended mode for 10HFTBch steel is the following: the end-of-rolling temperature is 850 °C, the beginning of accelerated cooling is 750 °C, the temperature of strip grinding into a roll is 600 °C.

Thus, in the case of the end of rolling, at 850 °C, the following mechanical properties were obtained, determined during the tensile test: tensile strength 540–560 MPa, impact strength 0.80–0.85 MJ/m^2, elongation 25–29% (Table 2.5). In the case of a decrease in the deformation temperature, an increase in the strength characteristics is achieved with a noticeable decrease in plasticity and toughness.

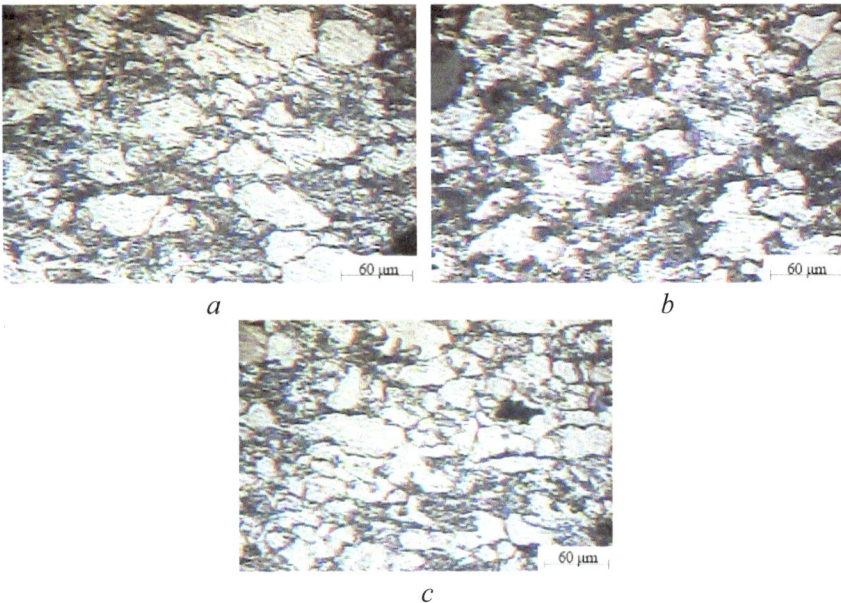

a   *b*

c

Fig. 2.45 Microstructure of deformed steel specimens at a deformation temperature of 800 °C (× 360): **a** zone of contact with the tool, I; **b** zone of periphery, II; **c** central zone of intensive deformation

Fig. 2.46 Macro structure of the deformed sample at 850 °C

2.5 Improvement of the Hot Strip Rolling Process

To test the workability of the developed integrated structure formation model for steel 10HFTBch, in hot rolling in the computer program Deform-3D, a series of experiments was conducted on physical modeling of energy-power rolling parameters in the finishing group of mill stands CBHRM 1680. Deform-3D allows correct calculations to be carried out in wide ranges of variation parameters. However, there are natural limitations of these ranges, which must be borne in mind when working with the program:

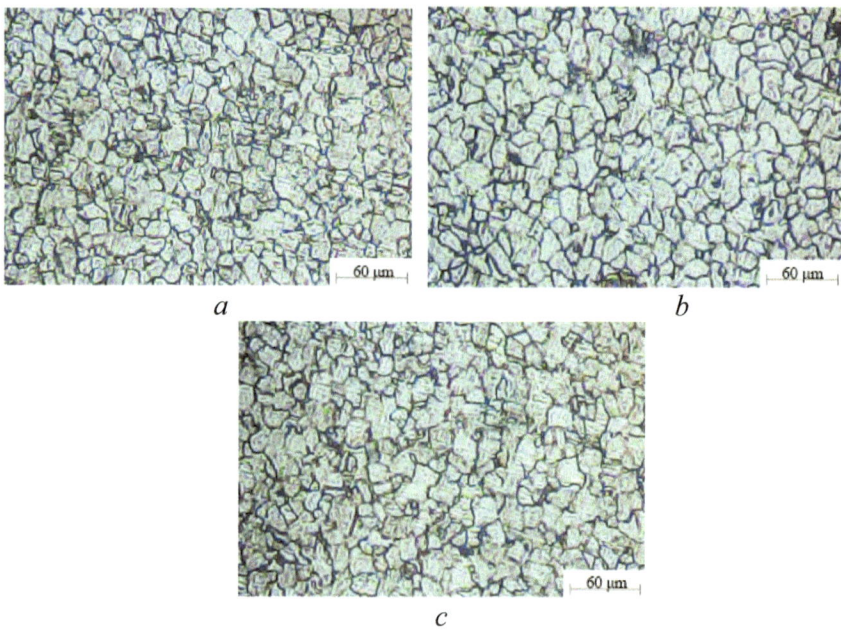

Fig. 2.47 Microstructure of deformed steel specimens at a deformation temperature of 850 °C ($\times$ 360): **a** zone of contact with the tool, I; **b** zone of periphery, II; **c** central zone of intensive deformation

Fig. 2.48 Macro structure of the deformed sample at 900 °C

- temperature $A_3 < T < 1500$ °C;
- in the degree of deformation $0 < \varepsilon < 1$;
- with respect to the strain rate $0.01 < u < 200$ s^{-1};
- the value of the original austenite grain $5 < \ < 350$ μm.

The parameters of the simulated rolling regimes, presented in Table 2.6, were chosen on the basis of the existing rolling mill conditions. The scheme of physical modeling of the deformation mode corresponds to the parameters given in Table 2.6, is shown in Fig. 2.52.

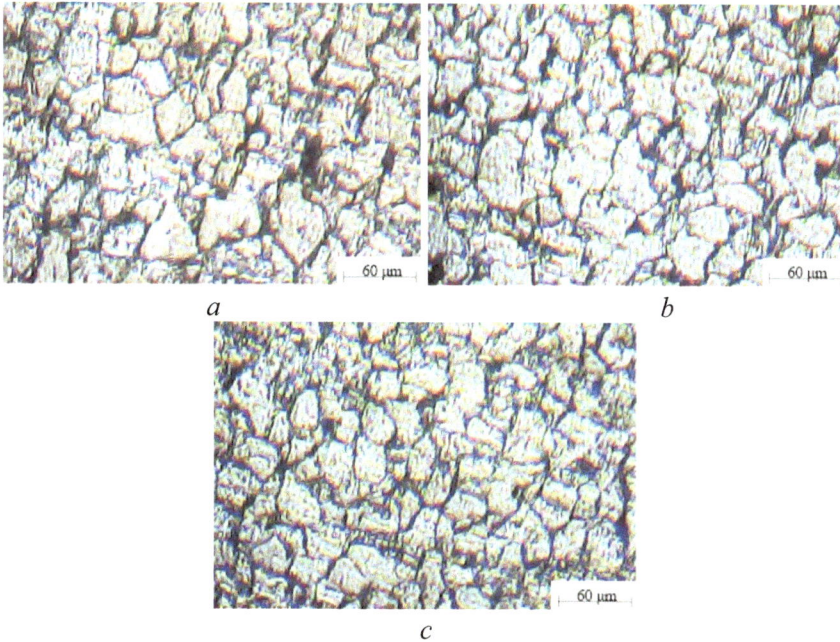

Fig. 2.49 Microstructure of deformed steel specimens at a deformation temperature of 900 °C (× 360): **a** zone of contact with the tool, I; **b** zone of periphery, II; **c** central zone of intensive deformation

Fig. 2.50 Macro structure of the deformed sample at 950 °C

When hot thin strips are rolled in the finishing groups of broadband mills, large contact normal stresses occur in the deformation centers of working stands, the magnitude of which is commensurable with the contact normal stresses in the deformation centers of cold rolling mills. As a consequence, elastic deformations in the contact of the strip and rolls have a significant effect on the energy-force parameters of the hot rolling process of thin strips. Therefore, as in the case of cold rolling, the calculation of the energy-strength parameters of broadband hot rolling mills ensuring a minimum discrepancy between the calculated and measured values, the rolling forces and the power of the main drive motors should be performed on the

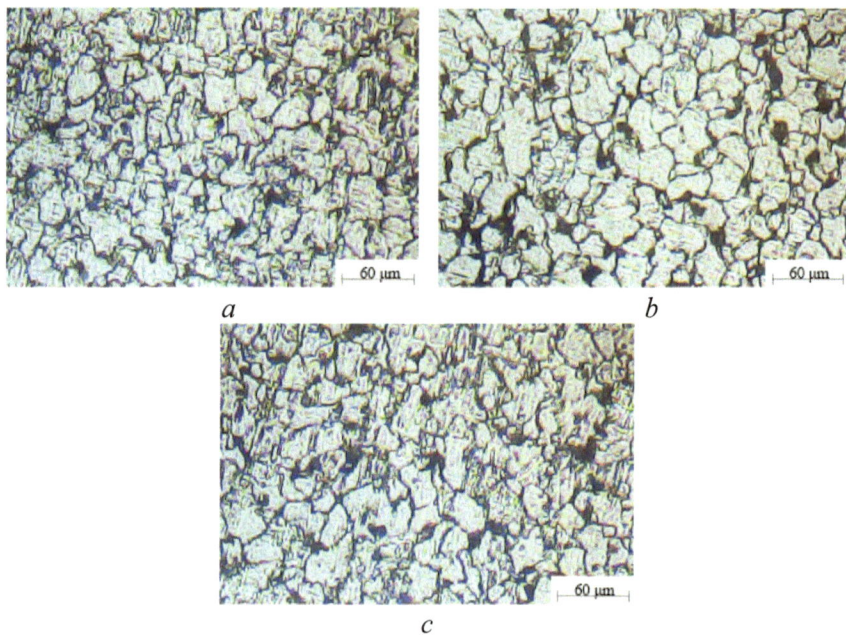

Fig. 2.51 Microstructure of deformed steel specimens at a deformation temperature of 950 °C (× 360): **a** zone of contact with the tool, I; **b** zone of periphery, II; **c** central zone of ihntensive deformation

Table 2.5 Chemical composition and mechanical properties of steels 10HFTBch, 10XM, Steel 15

Steel grade, patent	Content of basic alloying elements, in % mass, not less than					Mechanical properties, not less than[a]		
	C	Cr	Si	Mn	Other elements	σ_B,MPa	KCU (MJ/m²)	δ_5 (%)
10XM no. 81167	0.11	0.12	0.05	0.35	Mo—0.20	390	0.9	22
10HFTBch no.105341	0.10	0.10	0.30	0.32	Nb—0.11 Ti—0.12; V—0.12 B—0.001 P3M—0.005	550	0.8	28
Steel 15	0.15	0.2	0.06	0.4	Cu—0.25 Ni—0.25	360	0.8	30

[a] Mechanical properties are determined on the samples, after recrystallization treatment with 850 °C

Table 2.6 Modes of rolling strips on the mill CBHRM 1680 in the finishing group (thickness of the roll 29 mm mode 1 and 2, 28 mm—mode 3, 4)

Mode No.	The size section (mm)	Thickness of the strip along the stands h_i, mm, not less than					
		5	6	7	8	9	10
1	2.5 × 1000	15.4	9.3	5.6	3.9	2.91	2.5
2	3.0 × 1180	18.1	10.8	6.9	4.7	3.49	3.0
3	3.6 × 1200	17.3	11.3	7.5	5.3	4.17	3.6
4	3.8 × 1200	18.5	11.5	7.7	5.5	4.27	3.8

Fig. 2.52 Nomogram for selecting the hot rolling mode

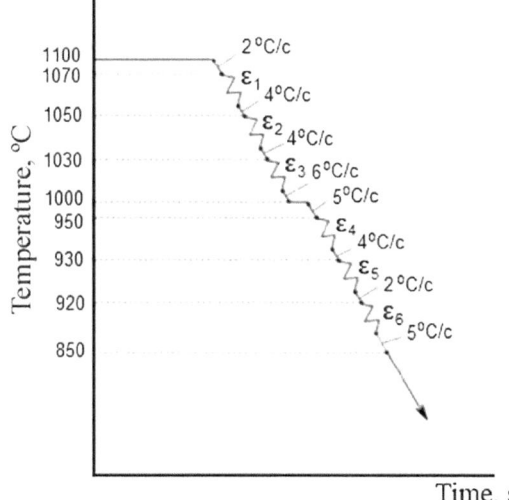

basis of the elastoplastic model of the stress–strain state of the strip in the deformation zone.

Using the method described above, using harmonic complex functions that include trigonometric dependencies on the coordinates of the deformation center, we calculated the normal contact stresses and energy-strength parameters of hot rolling, taking into account the results of this study (Sect. 2.3, 2.4), with the determination of the true resistance to deformation of steel 10HFTBch. To verify the reliability of the improved methodology, its software implementation was implemented.

When implemented, the following additional assumptions were used for the integral model:

(1) In the case of incomplete flow of dynamic recrystallization during deformation, the remaining, non-recrystallized part of the structure, recrystallizes according to the kinetics of static recrystallization;

(2) For multi-stage deformation, the effective grain size corresponds to the moment of $(i + 1)$-th deformation and is calculated by the rule of the mixture:

$$d^0_{\gamma,i+1} = d^{drx}_{\gamma,t}X_{drx,t} + d^{srx}_{\gamma,t}X_{srx,t} + \left(1 - X_{srx,t} - X_{drx,t}\right)d^0_{\gamma,t}; \qquad (2.19)$$

(3) After the i-th deformation, normal grain growth begins when the condition:

$$X_{srx,t} + X_{drx,t} \geq 0,95; \qquad (2.20)$$

(4) The resistance of plastic deformation at its (i + 1) -th stage is calculated by the formula:

$$\sigma_p = 0,000082\varepsilon^{0,524152}\exp\left(\frac{-0,000163}{\varepsilon}\right)$$
$$\exp(1,11363\varepsilon)(1+\varepsilon)^{-0,003363T}.$$
$$\cdot u^{-0,216846}u^{0,000336T}T^{2,97098}\exp(-0,004952T). \qquad (2.21)$$

(5) The value of the heat transfer coefficient for the strip in air is assumed to be 7.6 W/(m²K), the heat transfer coefficient of water flowing from the upper surface of the strip is 150 W/(m²K), the heat transfer coefficient of the jet is 13,000 W/(m²K). The water temperature in the calculations was assumed to be 30 °C, the working roll surface temperature was 65 °C, the blackness of the upper surface of the strip was 0.8.

The calculated diagram of the change in the mean mass temperature of the surface and the center is shown in Fig. 2.53.

The developed mathematical model for calculating the thermal state of a metal in the line of the final group of a continuous broadband mill made it possible to predict the energy-force parameters of rolling, the structure and properties of 10HFTBch steel after thermoplastic deformation.

Multiple recrystallization leads to an effective grinding of the austenite grain. The small grain size of austenite ensures the formation of a fine-grained ferritic structure, after the occurrence of ferritic transformation processes and leads to a decrease in the size of the pearlite colonies formed as a result of the pearlite reaction. Analysis of the temperature–velocity conditions for rolling strips 2.5–3.8 mm thick showed

Fig. 2.53 Results of calculating the temperature change of the strip in the finishing group of stands of the NSHCG 1680 mill

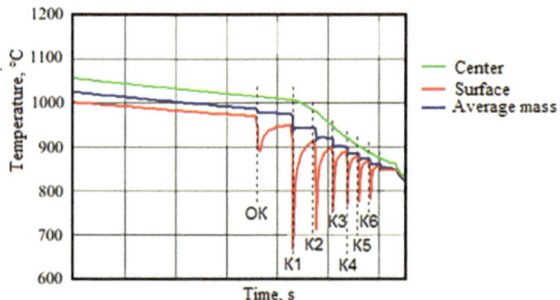

that the decomposition of austenite begins in the third-fourth cages of the finishing group and ends on a retractable roller table.

In Fig. 2.54 shows the distribution of normal contact stresses in finishing stands of a broadband mill, obtained by modeling in the Deform-3D module. It can be seen from the figure that the maximum values of normal contact stresses increase from 150 MPa in the fifth stand to 375 MPa in the tenth stand.

Thus, the results of calculating the energy-strength parameters showed that the value of the rolling force is within the permissible limits ($P_{max} = 13.2$ MN).

Fig. 2.54 Fields of distribution of normal stresses along the length of the deformation centers of the CBHRM 1680: **a** working stands no. 5; **b** working stand no. 10

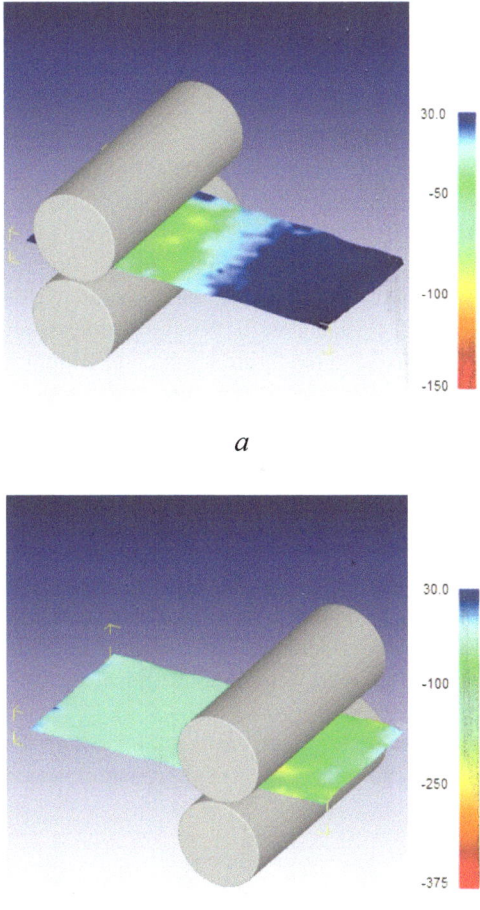

2.6 Industrial Realization of the Developed Thermoplastic Deformation Mode of 10HFTBch Steel

The effect of thermoplastic deformation of hot-rolled strips, from low-alloy steel 10HFTBch, on the microstructure and properties of rolled rolled products, was produced according to the serial rolling technology at the continuous broadband hot rolling mill 1680 (CBHRM 1680).

The assortment CBHRM 1680 includes hot-rolled sheets and strips with a thickness of 1.5 to 8.0 mm, a width of 1000–1200 mm, from low-carbon steels. Stainless and alloyed steel grades also have strips of thickness h > 3.5 mm and a width B < 1300 mm [28–30]. The starting material for the production of broadband steel is slabs of thickness 120–165 mm, weight $G = 5.5$ t and length Lc = 4.5–4.7 m, of which bands of thickness h < 2.5 mm (B = 1000 mm) [31]. Of slabs with a mass G < 15 tons and length L_{cl} = 9–9.3 m, bands of thickness h > 2.5 mm are rolled. The layout of the equipment of the CBHRM 1680 is shown in Fig. 2.55.

The heating temperature of slabs before rolling, depending on the chemical composition and purpose of the steel, is in most cases 1150–1300 °C. The slabs heated to the required temperature are supplied from the heating furnaces to the rolls of the vertical scavenger, the horizontal stand of the duo, the rough quarto stands No. 1–4. Raskat after the roughing group is fed to the intermediate rewinding device "Coilbox" and the flying scissors for trimming the ends of the roll.

The total reduction in the finishing group is 10–30%, of the total reduction throughout the mill. The applied compressions consistently decrease from the first (maximum reduction ~ 47%), to the last (up to 11–15%) stand (Table 2.7). Between the finishing stands are located loop holders, which ensure the tension of the strip during the rolling process; Device for descaling the scale, which are also used to reduce the temperature when rolling thick strips, as well as guide rails, and wiring.

For industrial approbation of the results of the work, thermoplastic deformation of hot-rolled strips was carried out, with a cross section dimension of 2.5 mm × 1000 mm, 3.0 mm × 1180 mm, 3.6 mm × 1200 mm, 3.8 mm × 1250 mm of low-alloy steel 10HFTBch, the chemical composition of which is indicated in Table. 2.1.

When determining the technological parameters of rolling, in the finishing group of stands, using different deformation-speed regimes, information was used for

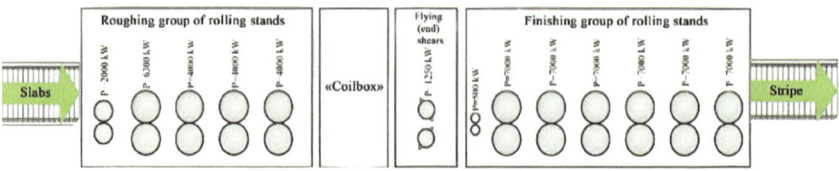

Fig. 2.55 Layout of the main equipment of the CBHRM 1680

Table 2.7 The magnitude of the deformation of the strips on the CBHRM 1680 in the finishing group (the thickness of the roll 29 mm mode 1 and 2, 28 mm—mode 3 and 4)

Mode No.	The size section (mm)	Relative reduction ε_i, %, not less than					
		5	6	7	8	9	10
1	2.5 × 1000	46.90	39.61	39.78	30.36	25.38	14.09
2	3.0 × 1180	37.59	40.33	36.11	31.88	25.74	14.04
3	3.6 × 1200	38.21	34.36	33.63	29.33	21.32	13.67
4	3.8 × 1250	33.93	37.84	33.04	28.57	22.36	1101

recording the control of the operation of drive motors and the values of stationary pyrometers installed in the slab line 1150-CBHRM 1680.

The following parameters were recorded:

- temperature of the end of rolling of ingots on slabbing;
- surface temperature of slabs before *CBHRM* 1680;
- the temperature of the rolling behind the stand No. 4;
- current of anchor circuits of drive motors No. 5–No. 10;
- speed of rotation of anchors of drive motors of stands No. 5–No. 10;
- voltage of anchor circuits of drive motors of stands No. 5-No. 6;
- moments on the shaft of the drive motors of stands No. 7–No. 10;
- end-of-rolling temperature;
- coiling temperature of the strips.

The thermoplastic deformation of sheet products from low-alloy steel 10HFTBch included caving of the roll before the last passes in finishing cages to temperatures close to the critical point Ar_3 (rolling-end temperature of 850 °C), deformation according to the regulated rolling regimes (Table 2.6), subsequent accelerated cooling, with a controlled rate of 15–20 °C/c and looking rolled into a roll at temperatures not higher than 580–600 °C.

Additional technological factors of hardening of steel during thermoplastic deformation, in comparison with heating for rolling, are: a decrease in the temperature of the end of the hot deformation below the recrystallization temperature of austenite and an increase in the degree of phase reduction at the last passes in finishing cages.

The obtained data on the actual loads, when rolling the research fusion strips, indicate that in the steady rolling process with constant speed, the values of the power of the drive motors in all cases do not reach the nominal value. The maximum value of the power is 5440 kW and is marked during rolling in the No. 7 rear end of the strip, with cross-sectional dimensions of 3.0 mm × 1180 mm, which is 0.78 of the rated engine power (7000 kW). In the noted case of rolling, when the value of the moment applied to the rolls is 1.01–1.07 times higher than the permissible values of their strength, they are explained by the reduced temperature of the metal strip, and by significant reduction in the No. 5 cage. The temperature margin of strength adopted in calculating the permissible loads of the rolls, provided their strength at the noted values of the moment. When rolling strips, the melts under investigation in subsequent finishing stands, the value of the moments fed to the rolls does not exceed the permissible values [32].

After thermoplastic deformation of sheet steel from 10HFTBch steel with thickness 2.5, 3.0, 3.5, and 3.8 mm in it fine-grained ferrite-pearlite structure of various morphology is formed. The size of ferritic grains, as a rule, does not exceed 8–12 μ. In individual grains of ferrite, a subgrain 3–4 μm in size is observed. Inside such subgrains, in turn, a thin polygonized structure with polygon sizes of 0.3–0.4 μm is revealed. The density of dislocations in ferrite is approximately 10^9 cm^{-2} (in the surface layers 1.5–2.0 times higher). Perlite has a classical lamellar structure. Its volume fraction increases by 2–4%, the interplanar distance in pearlite decreases to 0.2–0.3 μm. Often there are forms of the so-called "degenerate" perlite—thin discontinuous cementite plates in the ferrite matrix. The described three-layered structure of ferrite provides enhanced mechanical properties of thermoplastic deformed steel from steel 10HFTBch (Table 2.8). Table 2.8 shows the minimum test result of 3 samples per point. The actual rolling forces for the three tests are shown in Fig. 2.8.

Further, the microstructure, hardness and mechanical properties of welded joints of low-alloy steel 10HFTBch were investigated. The microstructure of the steel in the zone of thermal influence of the welded joint is a ferrite-pearlite structure of different dispersity: the grain size of the ferrite in the welded zone is 15–20 μm, in the zones of incomplete recrystallization and normalized—6 to 10 μm, while the base metal has 10–12 μm (Fig. 2.56).

The tensile strength of welded butt joints ($\sigma_s = 525$–540 N/mm^2) practically corresponded to the strength of the base metal. All tested samples were broken down on the base metal away from the welding site. The impact strength of samples with a U-notch along the welded zone is not much inferior to the toughness of the base metal (in the zone of thermal influence $KCU._{t.v.} = 0.80$–0.85 MJ/m^2, in the weld zone $KCU._{sh} = 0.75$–0.80 MJ/m^2, in the zone of the base metal $KCU_m = 0.90$–0.95 MJ/m^2) [33].

Table 2.8 Mechanical properties of thermoplastic deformed steel from steel 10HFTBch at a stretching

t (mm)	Sample[a]	σ	$\sigma_{0.01}$	σ_s	δ_5	δ_p	ψ
		MPa, not less than			%, not less than		
2.5	I	340	395	535	23	15	50
3.0	I	308	348	490	29	10	48
3.0	I	305	365	500	21	12	47
3.0	II	310	375	525	26	14	61
3.6	II	300	349	520	30	18	49
3.6	I	335	377	519	22	13	47
3.6	II	340	381	510	26	16	59
3.8	I	413	447	580	25	9	44

[a] The sample is cut in the transverse (I) and longitudinal (II) directions

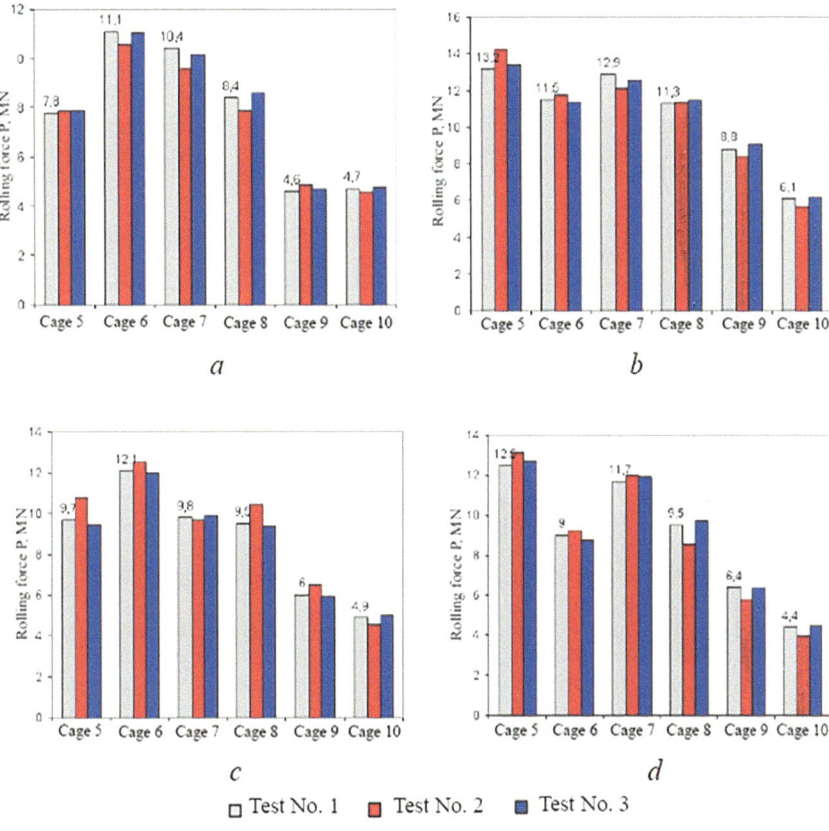

Fig. 2.56 Actual values of the rolling force when testing three samples per point: **a** 2.5 × 1000 mm, **b** 3.0 × 1180 mm, **c** 3.6 × 1200 mm, **d** 3.8 × 1250 mm

The developed technology of thermoplastic deformation of low-alloy steels of the type 10HFTBch provides, in sheet rolled products, a stable level of strength characteristics (tensile strength–540–560 MPa, with an average value of 550 MPa, plasticity—elongation—25 to 29%, with an average value of 27%), satisfying the requirements of technical documentation.

Thus, it becomes necessary to test high-strength low-alloy steel with the ultimate strength class K50 of a new doping composition, on the performance of wheels made of 10HFTBch steel.

References

1. Bolshakov V, Dolzhenkov Y, Dolzhenkov V (2002) Termycheskaia obrabotka staly y metalloprokata. Sich, Dnepropetrovsk (in Russian)

2. Golovanenko SA, Fonshtein NM (1984) Two-phase ferritic-martensitic steels. Met Sci Heat Treat 26(11):811–816
3. Kuziak R, Kawalla R, Waengler S (2008) Advanced high strength steels for automotive industry. Arch Civ Mech Eng 8(2):103–117. https://doi.org/10.1016/s1644-9665(12)60197-6
4. Spisak E, Hudak J, Tomas M (2010) Process formability of steels used in automotive industry. MM Sci J 2010(04):218–221. https://doi.org/10.17973/mmsj.2010_12_201019
5. Valberg HS (2010) Theory. In: Applied metal forming: including FEM analysis. Cambridge University Press, pp 53–76
6. Służalec A (2004) Theory of metal forming plasticity. Springer, Berlin, Heidelberg. https://doi.org/10.1007/978-3-662-10449-1
7. Shur EA, Kleshcheva II, Dudkina TP (1978) Fractures of steels with a heterogeneous structure. Met Sci Heat Treat 20(2):112–115. https://doi.org/10.1007/bf00670300
8. Malygin GA (2007) Plasticity and strength of micro- and nanocrystalline materials. Phys Solid State 49(6):1013–1033. https://doi.org/10.1134/s1063783407060017
9. Jonsson M (2006) An investigation of different strategies for thermo-mechanical rolling of structural steel heavy plates. ISIJ Int 46(8):1192–1199. https://doi.org/10.2355/isijinternational.46.1192
10. Vakulenko IA, Proydak SV (2014) The influence mechanism of ferrite grain size on strength stress at the fatigue of low-carbon steel. Sci Transp Prog Bull Dnipropetrovsk National Univ Railway Transp. 1(49):97–104. https://doi.org/10.15802/stp2014/22668
11. Vakulenko IA, Perkov ON, Razdobreev VG (2008) Mechanism of the effect of the ferrite grain size on the fatigue strength of a low-carbon steel. Russ Metall (Met) 3:229–231. https://doi.org/10.1134/s0036029508030087
12. Dyja H, Galkin A, Knapinski M, Ozhmegov K (2010) Plastometric test of zirconium alloy of system Zr–Nb. Hutnik Wiadomosci Hutnicze 5:207–209
13. Kolmogorov VL, Spevak LF, Churbayev RV (2008) On the technique used to determine plasticity margin in high-speed deformation under high pressure. Int J Mech Sci 50(4):676–682. https://doi.org/10.1016/j.ijmecsci.2008.01.002
14. Belodedenko S, Hrechanyi O, Vasilchenko T, Hrechana A, Izhevskyi Y (2023) Determination of the critical cyclic fracture toughness for the mode II in mixed fracture of structural steels. Forces Mech 13:100236. https://doi.org/10.1016/j.finmec.2023.100236
15. Belodedenko S, Hrechanyi O, Vasilchenko T, Baiul K, Hrechana A (2023) Development of a methodology for mechanical testing of steel samples for predicting the durability of vehicle wheel rims. Results Eng 18:101117. https://doi.org/10.1016/j.rineng.2023.101117
16. Sheyko S, Tsyganov V, Hrechanyi O, Vasilchenko T, Hrechana A (2024) Determination of the optimal temperature regime of plastic deformation of micro alloyed automo-bile wheel steels. Res Eng Struct Mater 10(1):331–339. https://doi.org/10.17515/resm2023.49me0428tn
17. Mishchenko VG, Sheyko SP (2014) Structural changes of multiphase low-carbon steel in deformation and heat treatment. Steel Transl 44(12):928–930. https://doi.org/10.3103/s0967091214120122
18. Mishenko V, Sheyko S, Bagriichuk O (2014) Analysis of the stress-strain state of the reactors in the process of titanium tetrachloride reduction. Austr J Sci Res 1(5):125–136
19. Chigirinskiy VV, Shejko SP, Dyja H, Knapinski M (2015) Experimental and theoretical analysis of stress state of plastic medium influence on structural transformations in low-alloy steels. Metallur Mining Ind 7(11):188–195
20. Spittel T, Hensel A, Spittel M (1983) Energy requirement and consumption in the metal deformation process. Hutnik Warszawa 50(3):87–95
21. Dyja H, Lesik L, Milenin A, Mroz S (2002) Theoretical and experimental analysis of stress and temperature distributions during the process of rolling bimetallic rods. J Mater Process Technol 125–126:731–735. https://doi.org/10.1016/s0924-0136(02)00378-3
22. Glowacki M, Mroz S, Lesik L (1999) Analiza podstawowych wzorow empirycznych obliczania parametrow energetyczno-sitowych procesu walcowania w wykrojach, Hutnik. Wiadomosci Hutnicze 11:523–529 (in Poland)

23. Sheyko S, Matiukhin A, Tsyganov V, Andreev A, Ben A, Kulabneva E (2021) Energy power parameter effect of hot rolling on the formation of the structure and properties of low-alloy steels. Eastern-Eur J Enterprise Technol 6(12(114)):20–26. https://doi.org/10.15587/1729-4061.2021.247269

24. Sheyko S, Mishchenko V, Matiukhin A, Bolsun O, Lavrinenkov A, Kulabneva E (2021) Universal equation of metal resistance dependence to deformation on conditions of thermo-plastic processing. Proceedings of the 30th Anniversary International Conference on Metallurgy and Materials. Brno, Czech Republic, TANGER Ltd., pp 329–334. https://doi.org/10.37904/metal.2021.4121

25. Mashekov SA, Mashekova AS, Urazbaeva RE, Kiyanbekova LR, Tussupkaliyeva EA, Smailova GA, Nurahmetova KK (2019) Special features of structure formation during rolling strips in the helical rolls and longitudinal wedge mill. News National Acad Sci Repub Kazakhstan 2(434):86–101. https://doi.org/10.32014/2019.2518-170x.42

26. Mashekov SA, Tussupkaliyeva EA, Akparova SA, Kiyanbekova LR, Nurakhmetova KK, Absadykov BN (2021) Influence of rolling modes on the anisotropy of sheet metals from carbon steel rolled on the longitudinal wedge mill (LWM) of a new design Metalurgija 60(1–2):125–128

27. Nemethova L, Kvačkaj T, Mišičko R (2009) Structural changes of C–Mn–Nb–V steel during the reheating. Acta Metallurgica Slovaca 3:173–179

28. Belodedenko SV, Hanush VI, Hrechanyi OM (2022) Experimental verification of the surviv-ability model under mixed I + II mode fracture for steels of rolling rolls. Struct Integrity 25:3–12. https://doi.org/10.1007/978-3-030-91847-7_1

29. Belodedenko S, Hanush V, Hrechanyi O (2022) Fatigue lifetime model under a complex loading with application of the amalgamating safety indices rule. Procedia Struct Integrity 36:182–189. https://doi.org/10.1016/j.prostr.2022.01.022

30. Belodedenko S, Hrechanyi O, Hanush V, Vlasov A (2022) Estimating the residual resource of basic structures using a model of fatigue durability under complex loading. Eastern-Eur J Enterprise Technol 3(1(117)):33–41. https://doi.org/10.15587/1729-4061.2022.257013

31. Belodedenko S, Grechany A, Yatsuba A (2018) Prediction of operability of the plate rolling rolls based on the mixed fracture mechanism. Eastern-Eur J Enterprise Technol 1, 7(91):4–11. https://doi.org/10.15587/1729-4061.2018.122818

32. Sheiko SP, Mishchenko VG, Matyukhin AY, Tsyganov VV, Tretyak VI (2021) Reserves for enhancing the mechanical performance of 10HFTBch low-perlite steel exposed to thermo-plastic processing in intercritical temperature ranges. Steel Transl 51(4):278–81. https://doi.org/10.3103/s0967091221040100

33. Sheyko S, Tsyganov V, Hrechanyi O, Vasilchenko T, Vlasov A (2023) Mechanical property investigation of welded joints in 10HFTBch steel for the automobile wheel production. Mater Lett: X 17:100181. https://doi.org/10.1016/j.mlblux.2023.100181

34. Mishenko V, Sheyko S (2014) Optimization of chemical composition of littlepearlitic steel of the special setting. Mod Sci Praha 2:68–76

Chapter 3
Modelling of Processes of Thermoplastic Processing of Intermetallic Alloys Under Non-stationary Temperature Conditions

Light alloys, based on titanium aluminide with TiAl phase, are currently considered as potential construction materials to use at temperature range 600–900 °C and they, as supposed, will find wide application in the near future. It has a unique complex of mechanical properties in comparison with traditional construction materials. It includes high specific strength and elasticity, which persist to high temperatures, high heat and oxidation resistance. These properties are due to the highly directional covalent component in the interatomic bond and the ordered atomic structure. At the same time, TiAl alloys are predominant in comparison with ceramic materials, owing to definite ductility and fracture toughness. The most promising application of TiAl alloys is in aircraft and space vehicles. The lightweight exterior panels with cellular filler and rigid thin-walled integral structures can be manufactured from these materials [1].

The main attention of the intermetallic γ-TiAl alloys developers in the last two decades was concentrated on achieving the optimal combination of mechanical properties by varying the microstructure from fully lamellar to duplex with varying grain size and plate thickness [2]. Depending on the alumina content alloys, based on γ-TiAl are divided into two groups: single-phase γ-alloys (50–52% Al) and two-phase $\gamma + \alpha_2$ alloys (44–49% Al). Through the obtaining technology with hot-deformation modes and heat treatment of biphasic alloys, three basic types of intermetallic structure are distinguished: lamellar, recrystallized (globular) and bimodal (duplex). Nowadays, three generations of industrial intermetallic alloys, based on γ-TiAl with different types of structure, are developed [3].

In our opinion, to give the TiAl alloy product the final properties, it is necessary to subject plastic deformation in the high-temperature phase region to obtain a plate structure. It provides the best combination of high-temperature properties—strength, creep resistance, with room ones—plasticity and fracture toughness. Apparently, plastic deformation can be effective not only for the production of fine-grained semi-finished products but also for controlling the parameters of the plate structure in TiAl alloys. In particular, it can be used for obtaining plate-like microstructures with a

S. Sheyko et al., *Thermoplastic Processing of Structural Metallic Materials*, SpringerBriefs in Materials, https://doi.org/10.1007/978-3-031-73896-8_3

small colony size and nanocrystalline interplanar spacing, which are of great interest [3].

The purpose of this work is to investigate the nanostructure formation in intermetalide γ-TiAl alloys by using a complex plastic deformation technology under non-stationary temperature conditions with niobium doping. Also in given work, the using of Hall-Patch model, the interrelation of the nanostructured quantities with strength characteristics is considered. It allows to obtain materials with increased plasticity by an optimal combination of mechanical properties over a wide range of temperatures.

3.1 Modelling of Deformation and Rheological Parameters for the Production of Intermetallic Alloys Under Conditions of Thermoplastic Treatment

Based on the results of works [4–6] and previous independent studies of the thermochemical pressing process (TCP process) [7, 8], an attempt was made to determine the main patterns of deformation and structure formation, to determine the ways and methods of controlling the processes of forming the structure and properties of pressed products. To solve the problem, the method of mathematical modeling was used, during the implementation of which the following main stages can be conventionally distinguished: idealization of the internal properties of the given process (object) and external influences (building a physical model); mathematical formulation of the behavior of a physical model (construction of a mathematical model); choosing a method for researching a mathematical model and conducting this research; analysis of the obtained mathematical result.

In the mathematical description of TCP-process it is necessary to take into account the thermokinetic characteristics of the process, the velocity of the reactant and its macroscopic density. Therefore, in addition to the kinetic equations for the formation of the intermetallic structure, the activation energy and chemical transformation, it is necessary to use the rheological equations used in describing the rhodinamic models, which allows us to carry out numerical calculations of the kinetic dependences of the basic parameters of the process of compressing the product of high-temperature synthesis—the temperature of synthesis, the completeness of chemical transformation, macroscopic the density of the product of synthesis, the level of elastic stresses in the product, the velocity of its melting point stycheskoy deformation and grain size finite product.

The starting material for SHS synthesis of the intermetallic compound TiAl is a powder mixture of nickel with aluminum, placed in the form of a molding in a closed mold. The powder compactor is warmed up to a given temperature and ignites in the mode of thermal explosion when the external pressure is applied, under the action of which the compression deforms. The plastic deformation ceases when the synthesis product is cooled to a temperature Tk, at which it loses ductility.

For a mathematical description of the process of extrusion of a high-temperature synthesis product, it is necessary to determine a system of equations that takes into account the distribution of the thermo-kinetic and rheological properties of the synthesis product in a mold and caliber. Assuming that the extrusion occurs in conditions of one-sided compression of the synthesis product in the absence of friction on the walls of the mold, we can write the initial equations:

1. Equation of continuity [9]:

$$\frac{\partial(\rho\rho_1)}{\partial t} + \frac{\partial(\rho\rho_1 V)}{\partial z} = 0 \tag{3.1}$$

2. Equation of motion [9]:

$$\rho\rho_1\left(\frac{\partial V}{\partial t} + V\frac{\partial V}{\partial z}\right) = \frac{\partial\sigma_{zz}}{\partial z} \tag{3.2}$$

with rheological correlations

$$\sigma_{zz} = \left(\frac{4}{3}\mu + \xi\right)\frac{\partial V}{\partial z}, \tag{3.3}$$

$$\sigma_{rr} = \sigma_{\theta\theta} = \left(-\frac{2}{3}\mu + \xi\right)\frac{\partial V}{\partial z}, \tag{3.4}$$

where ρ—relative density, ρ_1—density of the condensed phase, t—time, V—viscous flow velocity, z—axial coordinate, σ_{rr}, $\sigma_{\theta\theta}$, σ_{zz}—radial, tangential and axial stresses, μ, ξ—shear and bulk viscosity.

It is assumed that the distribution of relative density in the initial powder mixture is uniform:

$$\rho(z, 0) = \rho_0.$$

In the case where the density of the product SHS differs from the density of the original powder mixture

$$\rho_1 = \frac{\rho_c\rho_f}{\alpha\rho_c + (1 - \alpha)\rho_f},$$

where ρ_c—initial density of the mixture:

$$\rho_c = \frac{\rho_{Ti}\rho_{Al}}{\rho_{Ti}(1 - c_0) + \rho_{Al}c_0},$$

where ρ_{Al}—aluminum density; ρ_{Ti}—titanium density; ρ_f—the density of the reaction product; α—depth of chemical transformation of the intermetallic compound during the synthesis; c_0—relative mass concentration of titanium in the initial binary powder mixture.

The depth of chemical transformation is defined as

$$\frac{d\alpha}{dt} = f(\alpha)k_o\, exp\left(-\frac{E}{RT}\right),$$

where $f(\alpha)$—kinetic law; k_0—pre-exponential factor; E—activation energy of a chemical reaction.

To determine the conversion depth (α), the Johnson-Mel-Avrami-Kolmogorov model was used to estimate the kinetics of the formation of new phases and structural components. This model assumes that the appearance of a new phase occurs uniformly throughout the volume, the rate of appearance of a new phase does not depend on its already available quantity [14, 15]. The equation is written in the form:

$$\alpha(t) = 1 - exp\left(-Kt^n\right),$$

where K—is determined by the rate of growth of the phase in the volume and depends on the temperature and properties of the particular substance, n—parameter determined by the growth pattern of crystallites.

Different values of n correspond to different conditions for the formation and growth of embryos. If the cores are pre-formed and, therefore, they are all present from the very beginning, the transformation occurs only because of the 3-dimensional growth of the nuclei, then n is 3.

The parameter of crystallite growth rate K can be represented in the form:

$$K(T) \sim exp(-E_a/RT),$$

since the crystallization process is thermally activated.

The dependence of the shear and bulk viscosity of the synthesis product on its density is of the form [9]:

$$\mu(\rho) = \mu_1\rho^m, \quad \xi(\rho) = \frac{4}{3}\mu(\rho)\frac{\rho}{1-\rho} \tag{3.5}$$

where $\mu_1 = \mu_0\exp(U/RT)$—the viscosity of the non-flaking material basis, μ_0, U—physical constants, R—universal gas constant, T—temperature of the mixture, m—degree indicator.

3. Equation of thermal conductivity [6]:

$$c_1\rho\rho_1\left[\frac{\partial(\rho T_i)}{\partial t} + \frac{\partial(\rho V T_i)}{\partial z}\right] = \frac{\partial}{\partial z}\left[\lambda(p)\frac{\partial T_i}{\partial z}\right] + \rho\rho_1 Q\frac{\partial\alpha}{\partial t} - \frac{2\chi_i}{r_i}(T_i - T_0),$$

$$\tag{3.6}$$

where T_i T_i—the material temperature in the matrix (i = 1) and in the caliber (i = 2), $\lambda(\rho)$—dependence on the density of the heat conductivity of the material, χ_i—effective heat transfer coefficient, r_i—radius of the cross section of the matrix and the caliber, $c_1 = (1 - \alpha)c_s + \alpha c_{TiAl}$—heat capacity of the condensed phase; $c_s = c_{Ti}c_0 + c_{Al}(1 - c_0)$—heat capacity of the initial mixture, Q—thermal effect of intermetallide formation reaction TiAl, T_0—initial temperature.

The following assumptions were made to describe the process of compression of an intermetallic synthesis product in the mode of thermal explosion:

- the heating and cooling of the reacting powder system in the working space of the mold proceeds without a temperature gradient;
- the synthesis product is deformed in a homogeneous-stressed state;
- the heat sink from the mold can be neglected;
- the voltage at the upper boundary of the original powder compression in absolute value is equal to the compression force.

Thus, using the Eq. (3.1) and the relation (3.2), (3.3), the change in the density of the reactive in the mold of the powder system can be written as [10]:

$$\frac{\partial(\rho\rho_1)}{\partial t} = \frac{\rho\rho_1 N}{4/3\mu + \xi} \tag{3.7}$$

where N—the value of the applied pressure.

A equation of the thermal balance of a synthesis product, taking into account a number of assumptions, can take this form [10]:

$$c_1\rho\rho_1\frac{\partial T}{\partial t} = \rho\rho_1 Q_{TiAl}\frac{\partial\alpha}{\partial t} - \chi_1\frac{S}{V}(T - T_0), \tag{3.8}$$

where S—total area of the inner surface of the mold, V—volume of molds.

The Eqs. (3.7) and (3.8) allow us to quantify the parameters of the process of SHS pressing of the intermetallic compound synthesized under pressure. The process of forming a structure in a synthesized product under pressure is considered in the assumption that the initial grain size corresponds to the size of the initial particles of the refractory component (titanium), ie $D_0 = D_{Ti}$ (D_{Ti}—the diameter of the titanium particle).

The kinetics of grain growth as a result of the recrystallization of the synthesized intermetallic product is estimated from the equation [9]:

$$\frac{\partial D}{\partial t} = \frac{K(T_1)}{D^h}, \tag{3.9}$$

where D—the initial size (diameter) of the grain, $K = k_0 \exp(-E_a/RT)$—depends on the constant temperature, k_0—pre-exponential factor, E_a—energy of activation of grain growth, h—degree of magnitude close to 1.

The amount of deformation of the synthesized product during extrusion is determined from the equation:

$$\varepsilon \frac{r_1^2 - r_2^2}{r_1^2}. \tag{3.10}$$

Dependence of the grain size of the synthesized product on the degree of its deformation during extrusion is described by the empirical relations [11]:

$$D_\varepsilon = \frac{D}{\sqrt[3]{A\left(\dfrac{\varepsilon}{\varepsilon_{\kappa p}}\right)^2}}, \tag{3.11}$$

where $\varepsilon_{\kappa p}$–the degree of deformation at which the formation of the nucleation of recrystallization occurs ($\varepsilon_{\kappa p} \approx 0.1$), A—coefficient of form of intersection of the initial grain ($4\pi/3 < A < 6$).

A quantitative estimate of the grain size in the intermetallic product of the synthesis after extrusion can be carried out using the heat balance Eq. (3.7) and having carried out the derivative in time (3.8) in the derivative in the temperature.

Thus, after carrying out the necessary transformations, for the final grain size of the intermetallic product under the SWC compression, we can write [11]:

$$D_k = \sqrt{D_\varepsilon^2 + \frac{c\rho_0\rho_c r_2 RT_{a\partial}^2}{\chi_2 E_a(T_{a\partial} - T_0)} k_0 \exp\left(-\frac{E}{RT_{a\partial}}\right)}. \tag{3.12}$$

From Eq. (3.12) it is evident that the final grain size in the TCP-process product depends on the size of the grain of the product synthesized in the mold, the degree of deformation of the synthesized product during extrusion through the caliber, the adiabatic temperature of the synthesis of the extruded product and the speed of its cooling (depending on the temperature of the press shape, radius of its cross-section and coefficient of heat transfer of the synthesized product with the walls of the mold).

3.2 Numerical Analysis of the Influence of Technological Parameters on the Structure Formation Processes

Computer simulation of the hot-deformation processes of intermetallic γ-TiAl alloys is made using the software package Deform. The Deform program is a powerful system for modeling technological processes designed to analyze the three-dimensional behavior of the metal under various pressure processing processes. The program is based on the finite element method, one of the most well-known, reliable and currently used calculation methods. An automatic grid generator allows you to build an optimized finite element grid, thickening it in the most critical areas. In addition, the program provides important information on material flow and temperature distribution during the deformation process, which allows modeling a complete list of pressure processing processes and solving deformation and heat transfer problems. In solving the thermal deformation problem of compressing γ-TiAl alloys into the Deform program, the following output data were integrated:

$H_0 = 50$ mm, $r_1 = 25$ mm, $r_2 = 15$ mm, $T_{a\partial}$(TiAl) $= 1654$ K, $T_0 = 300$ K, $\rho_0 = 0.6$, $\rho_{Ti} = 4540$ kg/m^3, $\rho_{Al} = 2700$ kg/m^3, $\mu = 0.14$, $\rho_{TiAl} = 3800$ kg/m^3, $c_{Ti} = 540$ J/kg K, $c_{Al} = 929.5$ J/kg K, $c_{TiAl} = 600$ J/kg K, $Q_{TiAl} = 8.1 \times 10^3$ kJ/kg, E_a(TiAl) $= 79$ kJ/mol, $D_{Ti} = 100\,\mu$m.

In work [12], based on experimental research's methods of kinetic interaction in intermetallic alloys in SHS conditions, it was established that for obtaining γ-TiAl alloy the activation energy was nearly 79 kJ/mole and pre-exponential coefficient $k_0 = 7.2 \times 10^8$ s^{-1}.

In solving the thermoformation problem of compressing γ-TiAl alloys into the Deform program, the following data were integrated:

- the rheological properties of the γ-TiAl alloys $\sigma = f(\varepsilon, u, T)$, obtained experimentally on the Gleeble-3800 complex (Fig. 3.1) [13], which makes it possible to carry out numerical calculations of the kinetic dependences of the basic parameters of the process of compression of the product of high-temperature synthesis—the temperature of the system, the completeness of the chemical transformation, the macroscopic density of the synthesis product, the level of elastic stresses in the product, the speed of its plastic deformation and the grain size of the final product.
- parameters of the hydraulic press, according to the passport and the layout of the equipment;
- deformation and velocity (degree of deformation, displacement of the punch, etc.);
- temperature and temporal (thermophysical characteristics of the deformable and material of the technological instrument, coefficients of heat transfer, radiation, duration of pause, etc.).

To simulate the compression of the γ-TiAl alloy, the original finite element grid consisted of 100 elements grouped in a rectangle of 10 elements on one side. The sample in question was a cylinder 60 mm in diameter and 90 mm high.

The simulation results of the stress–strain state of TiAl alloys are presented in Fig. 3.2. The process of extrusion is characterized by a stress of comprehensive compression, which provides the material the best in these conditions plastic properties. Under the influence of compressive stresses, the material flows in the direction of the largest gradient of stresses—from the surface of the punch, where they have the maximum value, to the caliber of the matrix (Fig. 3.2b), where the normal stresses on the free surface of the tangent material are zero.

Comprehensive uneven compression provides the material with the highest ductility compared to other processes of metal treatment, but this feature of the process is manifested in extremely uneven deformations. In this case, only the compressive voltages acting continuously in the direction of extrusion from the maximum values to zero are not always in full volume of the deformed material. The presence of the difference between the intersections of the container and the

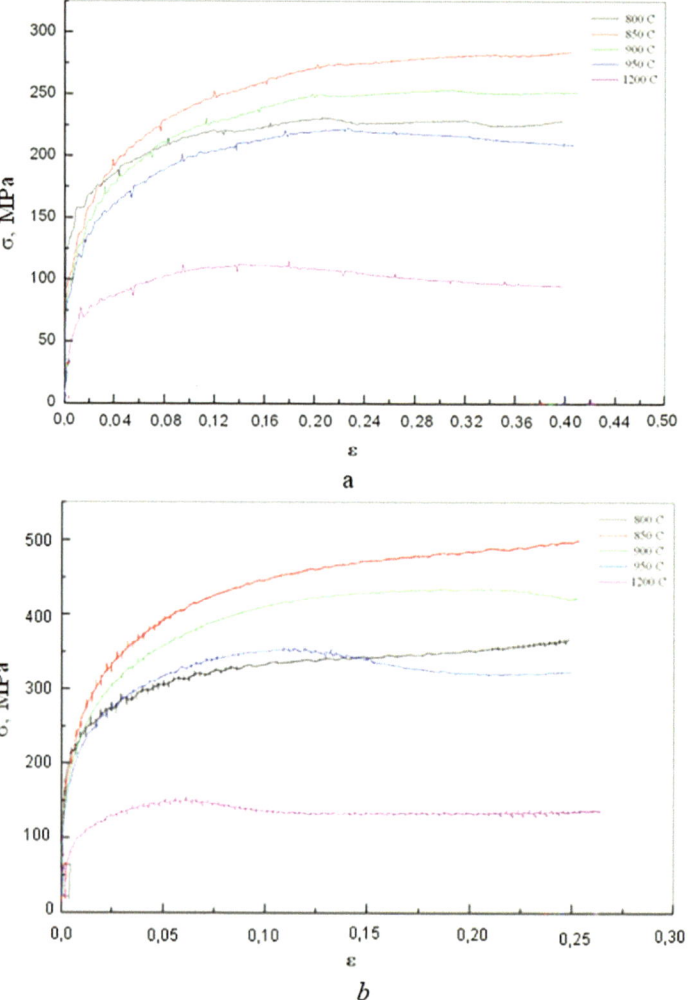

Fig. 3.1 Results of physical modeling of thermo-deformation treatment of the γ-TiAl alloy on the Gleeble-3800 complex: a is the strain rate, **a** $u = 0.1 \text{ s}^{-1}$, **b** $u = 1 \text{ s}^{-1}$

caliber of the matrix, the forces of contact friction and other factors leads to the fact that the particles of the material begin to move not only in the directions of the greatest deformation, but also in transverse directions. The latter contributes to the emergence of local (additional) stresses, the magnitude of different, direction and sign, and the emergence of tensile stresses. This is facilitated by the movement of material particles along trajectories of different lengths with velocity, change in the process of passage through different zones. The results of modeling the stress–strain state of TiAl alloys are shown in Fig. 3.3.

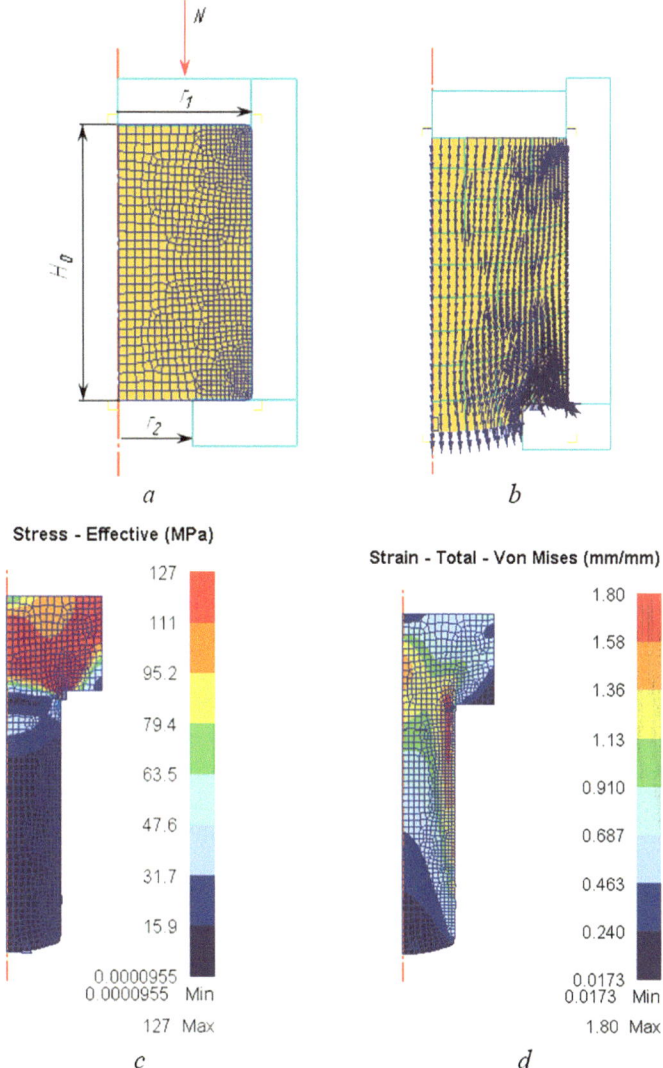

Fig. 3.2 Simulation of the process of SHS pressing of the intermetallic TiAl alloy in the program Deform: **a** the initial workpiece for calculation, **b** the direction of the tensile metal in the workpiece, **c** the pattern of the intensity distribution of stresses, **d** the intensity of deformation

In the conditions of the synchronization of thermal processes of the SHS and the dynamic compactation of the synthesis product, it is possible to obtain a compact intermetallic alloy with a highly dispersed structure, the size of which is much smaller than that of the alloys obtained by the methods of casting, sintering or shock-wave action on the synthesized product. Grinding of grain of intermetallic alloy in the process of its synthesis under pressure occurs as a result of plastic deformation of the

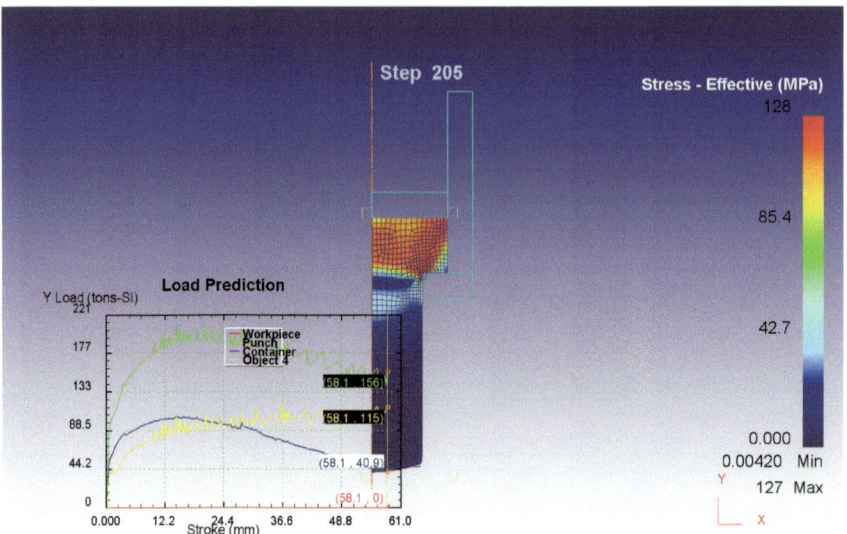

Fig. 3.3 Modeling of pressing processes of TiAl alloys

product of synthesis and high cooling rates (Fig. 3.4). High-temperature synthesis of the intermetallic compound γ-TiAl in a powder mixture of pure elements in the conditions of TCP-process at a thermal explosion at a minimum external pressure on the mixture allows obtaining an intermetallic synthesis product with an average grain size of ~ 30 μm. An increase in the degree of plastic deformation of the intermetallic product synthesized under pressure in the conditions of the TCP-process allows to reduce the size of the grain in the final product by an order of magnitude and even to form a sub-microcrystalline granular structure in the intermetallic alloy. The graphical interpretation of calculated results is shown on Fig. 3.4

Thus, a mathematical model aimed at obtaining γ-TiAl alloys with a given structure and properties is proposed and implemented, based on the use of data on the features of the physical modeling of the TCP-process process and the DEFORM software complex. High-temperature synthesis of intermetallic compound γ-TiAl in a powder mixture of pure elements in the conditions of TCP-process allows to obtain an intermetallic alloy with an average grain size of ~ 30 μ.

The experimental compacting curves and the results of modeling the pressing of TiAl alloys allow us to fix four clearly defined zones that determine the staging of the structure formation in TCP-process (Fig. 3.5).

The first stage of pressing—from the bulk density (the characteristic of the initial material) to the density of the pore level, is characterized mainly by structural deformation, particle re-packing, change in pore space. The consolidation of the powder mixture occurs due to a decrease in the volume of air inclusions in the material and the closure of macropores.

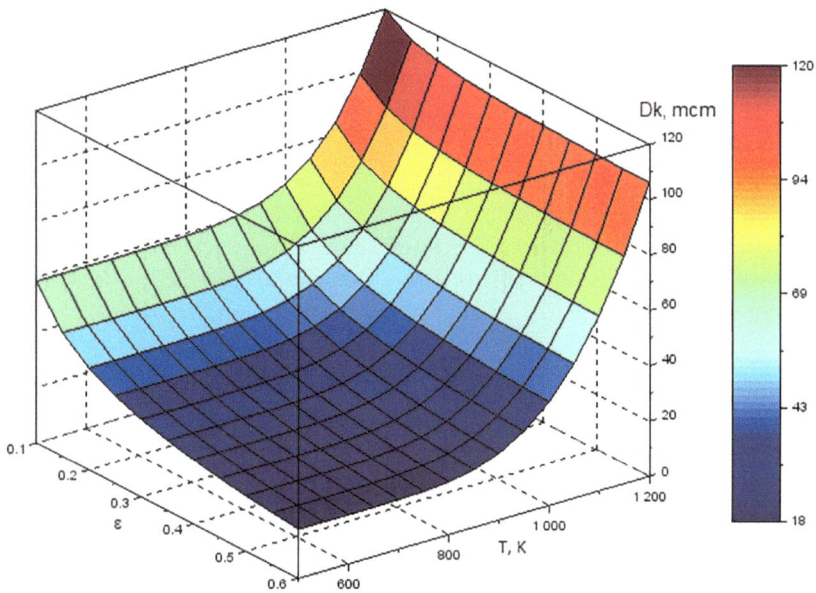

Fig. 3.4 The dependence of grain size of TiAl intermetallide on deformation and temperature degree

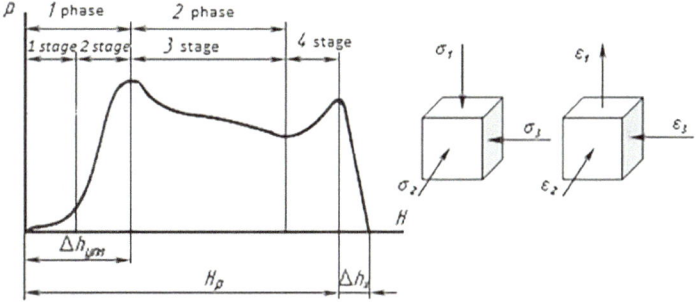

Fig. 3.5 Mechanical deformation scheme and loading schedule of TCP-process

The second stage, the stage of thermal autoignition, shows an abrupt increase in the relative density, which indicates a certain self-tightening of the γ-TiAl alloy during the synthesis, apparently under the action of surface tension forces. The initial stage of the formation of titanium aluminide occurs.

The third stage—compaction is characterized by structural deformation. The threshold density of the stage is 83–95%. In a system containing 39.6 wt% Al, the layer previously formed limits the movement of aluminum atoms to the titanium material. At the same time, the TiAl$_3$ layer builds up, which leads to an abnormal aluminum mass and the subsequent formation of titanium monoaluminide.

The fourth stage—the final stage of the structure formation will be the alignment of the composition of the intermetallic layers, primarily due to the recrystallization of TiAl$_3$ in TiAl and the secondary structure formation of Ti$_3$Al as a result of dissolution of the inner titanium core. In the last stage of compaction, the compact density reaches 98–99% of theoretical density. The residual porosity of the intermetallic compound is 1–2%.

Thus, preliminary research method aimed at obtaining γ-TiAl alloys with a given structure and properties based on the use of data on the physical modeling of thermal deformation processing on the Gleeble 3800 complex and the DEFORM software is proposed and implemented.

3.3   Effect of Hot Deformation on the Structure Formation and Properties of Intermetallic Alloys

The investigated materials include elemental powders of 99.8% pure titanium with an average particle size of 100 μm, 99.6% pure aluminum with an average particle size of 70 μm and 99.8% pure niobium with an average particle size of 100 μm. The powders were preliminarily mixed at a stoichiometric ratio for two systems in a Uniball 5 mill using any balls:

   Ti–Al: Ti—48%, Al—52%;
   Ti–Al–Nb: Ti-42%, Al-50%, Nb—8%.

The obtained mixture was compressed in a mold on hydraulic press with pressure of 300 MPa. The combustion process was carried out in the special furnace with protective atmosphere under normal conditions. The compaction of the synthesized samples was carried out at pressure of 100 MPa.

To evaluate the microstructure properties of the intermetallic compound, the rectangular parallelepiped samples were cut by size 5 mm × 10 mm × 10 mm.

The microstructure of the obtained alloys was researched with scanning electron microscope "SUPRA 40 WDS".

To calculate limit stress of obtained alloys the Hall-Patch model was used [5–7].

The scientific novelty of the work is that the high-temperature synthesis of the intermetallic compound γ-TiAl in a powder mixture of pure elements, under the conditions of the TCP-process, at a thermal explosion at a minimum external pressure on the mixture allows to obtain intermetallic synthesis product with an average grain size of 30 μm.

Based on the results of our earlier works, related with TCP-process [14, 15], it is necessary to establish the basic laws of deformation and structure of intermetallic alloys with unsteady temperature conditions.

The analysis of calculations has shown that in conditions of synchronized thermal SHS processes and dynamic compaction product synthesis, it is possible to obtain a compact intermetallic alloy of fine structure, which has grain size that much less

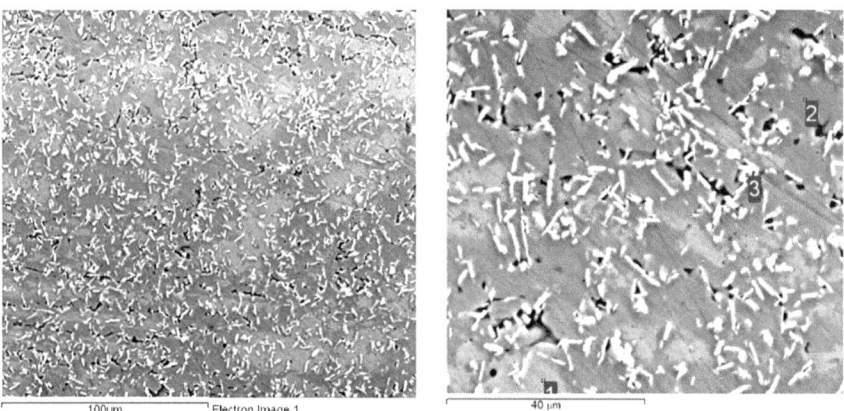

Fig. 3.6 The microstructure of obtained two-phase (γ/α_2) TiAl alloy

than the alloys, obtained by casting, sintering or shock-wave action. It is found that high-temperature synthesis of intermetallic alloy TiAl powder in a mixture of pure elements in terms of TCP-process with thermal explosion provides a synthesis of intermetallic product with an average grain size ~ 30 μ. The increasing of plastic deformation degree will significantly reduce the grain size in the final product and more over will form intermetallic alloy with sub-microcrystal granular structure.

The fine-grained two-phase duplex structure of alloys has the best plasticity but at the same time, another not less important characteristic—the viscosity of the alloy is reduced. The optimum option is to obtain alloys with a fully lamellar two-phase (γ/α_2) structure with a certain amount of γ- and α_2-phases in the alloy (Fig. 3.6).

The results of X-ray diffraction analysis showed that in the synthesized state, the γ-TiAl alloy consists of two phases: TiAl (γ-phase) and Ti_3Al (α_2-phase). In a given alloy, the volume fraction of the α_2-phase relative to the γ-phase is about 20%. The diffraction pattern of the synthesized γ-TiAl alloy is shown in Fig. 3.7. The X-ray diffraction analysis method found that on the diffractograms of the samples after the synthesis there are peaks of γ-TiAl (interplanar distances $d = 2.2063, 1.9120, 1.2811, 1.1777, 1.1468$ Å) and weak reflections of Ti_3Al peaks ($d = 2.1036, 1.3902$ Å). That is, the phase composition of the investigated synthesized γ-TiAl alloys is similar to that found earlier in papers [16, 17].

Metallographic studies have shown that the synthesis of the alloy formed a two-phase structure. In the alloy, there are single micropores, the presence of large pores and cracks are not detected. The microhardness of the alloy is HV 3000–4000 MPa. The results of microanalysis revealed a uniform and fine distribution of titanium aluminide TiAl. According to the microstructures, the TiAl system was predicted to be two-phase: TiAl (γ-phase) gray and Ti_3Al (α_2-phase) white. In addition, in the structure of the alloy, on the background of a two-phase structure, disperse light inclusions of various forms are formed, which are evenly distributed in the volume of the matrix and have an increased content of titanium.

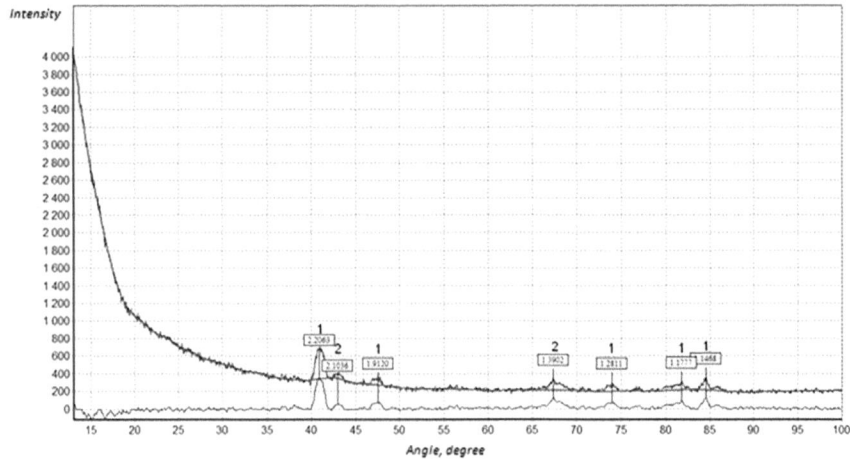

Fig. 3.7 The diffraction pattern of the synthesized γ-TiAl alloy: 1—the phase of γ-TiAl; 2—α_2-Ti$_3$Al

The chemical composition of the γ-TiAl alloy in different areas of the surface microshields was determined using the microrentgenospectral analysis (Fig. 3.8). The content of components was determined in atomic and mass percentages. As a result of the quantitative analysis it was found that the matrix (gray area) of the γ-TiAl alloy has a composition in mass percentage: 42.13% Al and 57.52% Ti, which corresponds to the intermetallic phase of TiAl (γ-phase) (Fig. 3.8a, spectrum 3). Extruded, white in the region, according to the results of microrentense spectral analysis (Fig. 3.8a, spectrum 7) are intermetallics of α_2-Ti$_3$Al composition, with elements content in mass percentages: 22.62% Al and 77.38%.

Near these inclusions, and in some cases and within them, the dispersed phases are found. Indications of local chemical analysis (Fig. 3.8a, Spectrum 1) allowed them to be identified as α_2-Ti$_3$Al intermetallic substances containing 28.77% Al and 71.23% Ti. This is consistent with the results obtained in [10].

To clarify the data, micro-X-ray spectral analysis was performed along the line (Fig. 3.8b). In the left part of the scanning trajectory are marked peaks of aluminum, which confirms the crystallization of the intermetallic phase in the alloy in the form of monoaluminidum titanium γ-TiAl. Further motion in the trajectory of scanning (Fig. 3.8c) in the region of the extended form shows the increase in the content of titanium and the reduction of the content of aluminum (Fig. 3.8d). This is consistent with the results of the local analysis and indicates the formation of the intermetal phase α_2-Ti$_3$Al.

The results of the microrentense-spectral analysis allowed to prove the production of a two-phase structure in the γ-TiAl alloy with the intermetallic phases γ-TiAl and α_2-Ti$_3$Al. It was also found that due to the high temperature of synthesis, the process of self-cleaning of the product from impurities takes place in the IMS, which minimizes the probability of the appearance of impurity atoms. The determination

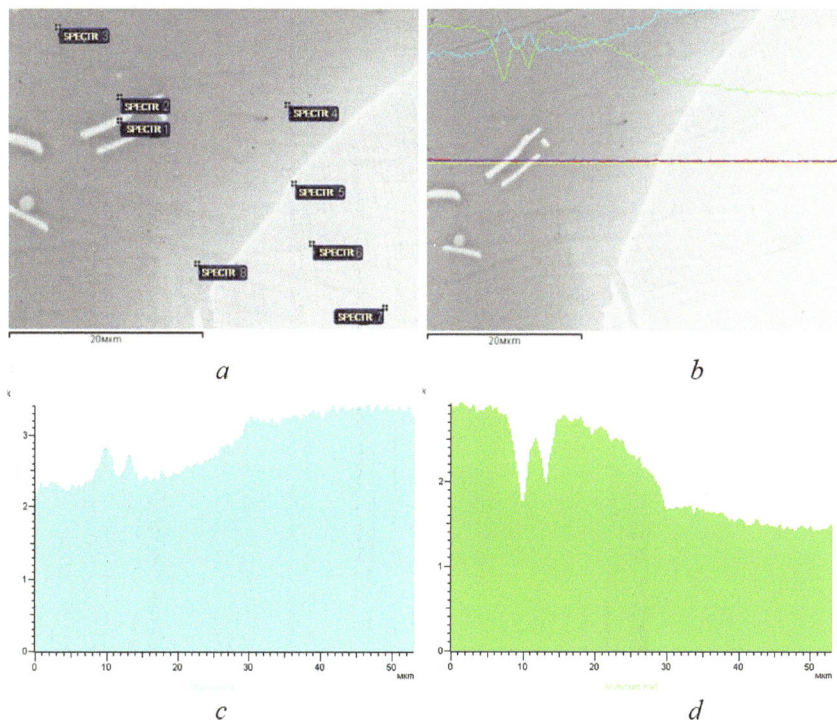

Fig. 3.8 Results of microrentense-spatial analysis of γ-TiAl alloy: **a** location of determination of local chemical analysis of alloy; **b** change in the intensity of radiation in motion along the line; **c** distribution of titanium; **d** distribution of aluminum

of the SUPRA 40WDS microscope in the SHS alloy of oxygen, nitrogen and other impurities showed that they are missing or not exceeding one thousandth of a percent.

Doping with a refractory Nb is the most important for the application of products where high temperature strength and oxidation resistance, as well as room temperature ductility, are critical. The niobium addition has a significant effect on the volume of α_2-phase fraction and average interlam distance. Thus, doping of γ-TiAl alloys with niobium (7–8% by weight) and an increasing of plastic deformation degree in conditions of extrusion at a load of 100 MPa allows to significantly reduce the grain size in the final product (to 0.2–0.3 μm). It leads to submicrocrystalline granular structure formation in the intermetallic alloy.

The microstructure analyze of synthesized Ti–Al–Nb alloy has shown that the formation features of thinner structures is in increased content of the β-stabilizing element. As a result, a thin composite texture is formed consisting of parallel alternating lamellae of two different crystalline phases: tetragonal γ-phase (TiAl) and hexagonal α_2-phase (Ti$_3$Al) (Fig. 3.9). Thus, a two-level structure is formed: each polycrystalline α-grain forms a bounded lamellar colony consisting of thin lamellas with an intermolecular distance of 500 nm.

Fig. 3.9 The microstructure of obtained Ti–Al–Nb alloy

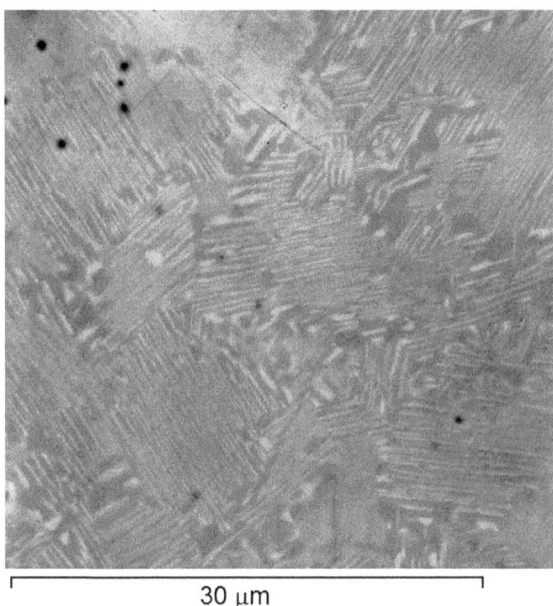

30 μm

The further calculation is aimed at studying the strength evaluation of alloys according to the Hall-Patch model. The given theory gives the following dependence of short-term limit stress (σ) on the grain size (d) and lamellae thickness (λ) during faze segregation (Fig. 3.10).

$$\sigma = \sigma_0 + \frac{k_d}{\sqrt{d}} + \frac{k_\lambda}{\sqrt{\lambda}}$$

where: σ_0 is the yield stress of non-textured material, k_d and k_λ are Hall-Patch constants of grain and lamellae accordingly.

The mechanical properties of the two-level structure can be improved upon the transition of parameters of structural-phase segregation d and λ from micron to nano-dimensional level. In addition to the thermal conditions, the segregation parameters are optimized by doping affects on values of the coefficients k_d, k_λ. The best result is achieved with the joint action of these factors.

The limit strength of fully lamellar intermetallic alloys with colony sizes ranging from d = 10–50 μm and lamellar spacings from λ = 100–500 nm was studied. As noted in work [18], an apparent Hall–Petch constant of $k_d = 2.1$ MPa·m$^{1/2}$ was previously determined when the limit stresses were ranked against the colony size. It has been recognized that the lamellar spacing is proportional to grain size d and limit stress is governed by λ. In order to make this idea quantitative, the authors proposed an analytical model for the limit stress of lamellar material that involves two Hall–Petch constants, which account for the effects of the grain boundaries ($k_d = 0.91$ MPa m$^{1/2}$) and the lamellae interfaces ($k_\lambda = 0.45$ MPa m$^{1/2}$).

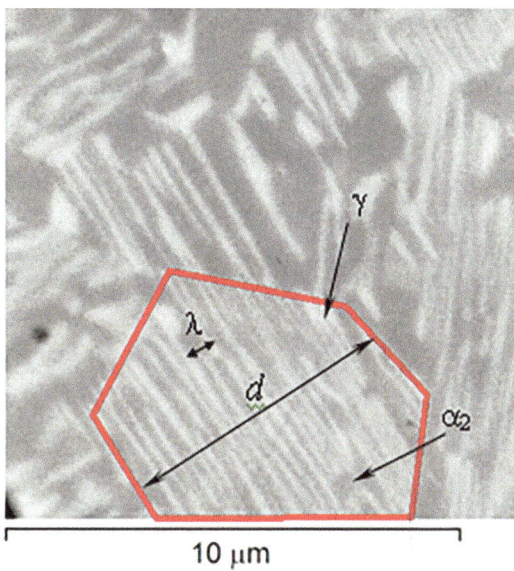

Fig. 3.10 The grain size d and lamellar colonies distance λ in the Ti–Al–Nb system

10 μm

The results intermetallic alloys strength calculations are presented on Fig. 3.11. Calculations, which were carried out using the Hall–Petch model [19], show the possibility of obtaining structural components with a size of 10–12 μm with a tensile strength of more than 1800 MPa for Ti-Al-Nb and 800 MPa for Ti-Al. Analysis of literature data showed that the strength limits values of these materials are close or even exceeding the obtained values for a certain size of structural components, which indicates the reliability of the results. It is important to note that the plastic anisotropy of the lamellar morphology can also become evident in the strength properties, which will be the subject of our further research.

Fig. 3.11 The dependence of stress limit σ on grain size d for Ti–Al–Nb (1) and Ti–Al (2)

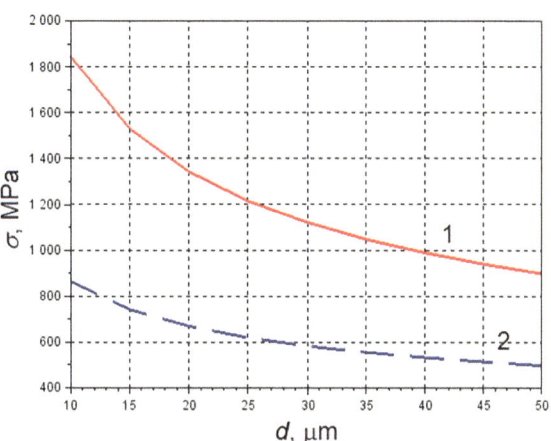

It is found that high-temperature synthesis of intermetallic alloy TiAl powder in a mixture of pure elements in terms of TCP-process with thermal explosion provides a synthesis of intermetallic product with an average grain size ~ 30 μ. The increasing of plastic deformation degree will significantly reduce the grain size in the final product and even will form intermetallic alloy with sub-microcrystal granular structure.

Microstructure analize has shown that doping of γ-TiAl alloys with niobium (7–8% by weight) and an increasing of plastic deformation degree in conditions of extrusion at a load of 100 MPa allows to significantly reduce the grain size in the final product (to 10–12 μm) and form two-level structure with nanolamellar colony with distance up to 500 nm.

Calculations, carried out using the Hall–Petch model, show that obtained Ti–Al–Nb alloy with lamellae nanostructure has limit strength up to 1800 MPa which in 3 times more than in Ti–Al alloy. Analysis of literature data show that the strength limits values of these materials are close or even exceeding the obtained values for a certain size of structural components, which indicates the reliability of the results.

3.4 Development of Intermetallic Alloys with High Level of Physico-Mechanical and Operational Properties

He main attention of developers of intermetallic γ-TiAl alloys in the last two decades has been concentrated on achieving the optimal combination of mechanical properties by varying the microstructure from completely lamellar to duplex with different grain sizes and plate thicknesses [20]. Depending on the aluminum content, γ-TiAl-based alloys are divided into two groups: single-phase γ-alloys (50–52% Al) and biphase γ + α_2 alloys (44–49% Al). Three main types of intermetallic structure based on titanium aluminides: lamellar (lamellar), recrystallized (globular) and bimodal (duplex)) are distinguished depending on the technology for obtaining blanks, hot deformation modes and thermal processing of two-phase alloys. In the foreign literature there is a classification into four types of structure: near-gamma, duplex, nearly-lamellar, fullylamellar. To date, there is no universal γ-TiAl alloy, the characteristics of which would fully satisfy the requirements of the aerospace and aerospace engineering developers for the whole spectrum of operational properties. Optimization of the chemical composition and microstructure of these materials led to the generation of their three generations (composition in atomic percent) [1–3].

1st generation—Ti-48Al-1 V-0.3C.

2nd generation—Ti–(45–48) Al–(1–3)X—(2–5)Y—(<1)Z.

where: X = Cr, Mn, V; Y = Nb, Ta, W, Mo; Z = Si, B, C.

3rd generation—Ti—(45–47) Al—(5–10) Nb—(<1) B, C.

Industrial γ-TiAl alloys of the second generation contain at least one X-element and one ϒ-element, which increase the resistance to oxidation and creep. Like high-temperature nickel superalloys, they can contain up to eight alloying elements. These alloys have good workability, satisfactory strength properties, elongation at a stretching of 1–3% at room temperature, fracture toughness from 10 to 25 MPa/m [20]. However, according to the creep characteristics, their use is limited to 700 °C, especially during long-term operation. At temperatures above 700 °C, insufficient oxidation resistance can also be affected.

Alloys of the third generation are developed in order to increase their operating temperatures. The works are conducted in two directions: (1) on the basis of alloys with a high content of niobium; (2) development of dispersion-hardened alloys [21, 22]. TiAl-based alloys with niobium content from 5 to 10% and small additives B and C are TNB [23]. These alloys have higher strength and oxidation resistance compared to second-generation alloys. In the opinion of the author [24], TiAl-based alloys can have an acceptable creep resistance up to a temperature of 750 °C with a duplex (lamellar-granular) structure, and up to 950 °C with a lamellar structure.

The latest trend in the development of technology for high-temperature γ-intermetallic compounds based on TiAl is associated with their special microstructuring, i.e. the decrease in both the mean size of the primary polycrystalline grain and the thickness of the γ- and α_2-phase lamellae after post-crystallization solid-phase transformations occurring in accordance with the alloy state diagram of a specific chemical composition [20, 22, 25–27]. Innovative γ-TiAl alloys (TNM-like alloys) contain 42–46 at.% Aluminum, and as ligatures in the sum up to 10 at.% Transition metals. In addition to the obligatory Nb, such β-stabilizers as Mo, Ta, Zr, Cr, W, V can be used [28].

At the same time, to date, there are no strict analytical relationships connecting the parameters of the structure of TiAl-intermetallides and their composition with the mechanical strength characteristics of the material. The development of materials science in this area is at the level of empirical research, taking into account the qualitative laws, therefore when planning the work it is possible to be guided only by independent structural, composite and strength studies of samples in order to obtain empirical data. The most promising microstructure of the cast alloy is characterized by the presence of ultra-fine equiaxed grain-colonies completely filled with laminated TiAl and Ti_3Al-phase lamellae. This microstructure is characteristic of alloys based on Ti–Al–Nb.

The mechanical properties were determined on standard discontinuous specimens in accordance with [29] on a tearing machine MUP-20 at a load of 5 tons and the speed of movement of the active gripper 2.5 mm/min. During the test of the sample, a tensile diagram was recorded, fixing the relationship between the force acting on the sample P and the deformation induced by it Δl.

Optimization parameters:

Y_1—ultimate strength (σ_s), MPa → max;
Y_2—relative elongation (δ), %. > 4.

As independent variables, the content in the niobium alloy (X_1), the content of molybdenum (X_2), the content of the chromium mixture (X_3) were chosen. As the initial components, pure powders of titanium, aluminum, niobium, molybdenum and chromium were used. Dispersion of powders was 50–100 μ. The batch preparation scheme included dosage, mixing, mold filling, pressing and heat treatment.

Table 3.1 Factors studied

Characteristic	Factors		
	Nb (% wt)	Mo (% wt)	Cr (% wt)
Code	X1	X2	X3
Basic level	8	3	2
Variation interval	4	2	1
Lower level	4	1	1
Top level	12	5	3

To obtain the quadratic coefficients of the regression equation, an orthogonal plan of the second order with the kernel 2^3 is used [30]. The calculated levels of variation intervals, the nature of their variation and the coding scheme are presented in Table 3.1. Niobium, applied at high concentrations, increases the amount of the α_2 phase in alloys, grinds the microstructure, increases the oxidation resistance. Chromium reduces the energy of the defects of the package, which leads to an increase in ductility at room temperature as a result of increasing the tendency of alloys to twinning. Molybdenum, which has a high β-stabilizing activity, grinds the grain.

As a result of regression analysis, according to the procedure considered in [30], equations were obtained showing the dependence of the mechanical properties of the γ-TiAl alloy on the content of alloying elements:

$$Y_1 = 1198, 22 + 20X_1 + 12X_2 + 7X_3 - 147, 78X_1^2$$
$$- 37, 78X_2^2 - 12.78X_3^3 - 5X_1X_2$$
$$- 2, 5X_1X_3 - 2, 5X_2X_3; \tag{3.13}$$

$$Y_2 = 3, 89 + 1, 02X_1 + 0, 54X_2 - 0, 08X_3$$
$$- 0, 97X_1^2 - 0, 37X_2^2 + 0, 33X_3^3 - 0, 16X_1X_2$$
$$+ 0, 04X_1X_3 - 0, 01X_2X_3 \tag{3.14}$$

Checking the significance of the regression coefficients by the Student's test and evaluating the adequacy of the model using the Fisher criterion are presented in Table 3.2.

Table 3.2 Checking the results of regression analysis for relevance and relevance

Parameter	Response functions	
	Y1	Y2
Δb	10.14	0.10
t-test	2.78	2.78
F-test	6.09 > 3.69	6.09 > 6.02

The coefficients, the absolute value of which is equal to the confidence interval Δb or more, should be considered statistically significant. Statistically insignificant coefficients (in this case, b_{12}, b_{13}, b_{23}) can be excluded from models.

Checking the adequacy of the models shows that they can be used to predict the values of the response functions for any values of the factors between the upper and lower levels. For this, it is advisable to go over to natural variables using the translation formula, which is presented in the following form [31]:

$$X_{ij}^k = \frac{X_{ij}^n - X_{ij}^o}{\Delta_i}, \tag{3.15}$$

where X_{ij}^k—coded value of the studied i-th factor in the j-th equation; X_{ij}^n—natural value of the studied i-th factor in the j-th equation; X_{ij}^0—value of the studied i-th factor in the j-th equation at the ground level; Δ_i—value of the variation interval of the i-factor studied.

By replacing the variables Xi in Eqs. (3.13)–(3.14) with the right-hand side of Eq. (3.15) and the subsequent reduction of similar ones, we obtain natural equations describing the effect of the content of alloying elements on the mechanical properties of γ-TiAl alloys:

$$\sigma_s = 413 + 152.78\text{Nb} + 62.67\text{Mo} + 51.11\text{Cr}$$
$$-9.23\text{Nb}^2 - 9.44\text{Mo}^2 - 12.78\text{Cr}^2 \tag{3.16}$$

$$\delta = -2.8 + 1.28\text{Nb} + 0.98\text{Mo} - 1.33\text{Cr} - 0.06\text{Nb}^2$$
$$-0.1\text{Mo}^2 + 0.33\text{Cr}^2 - 0.02\text{NbMo}. \tag{3.17}$$

As can be seen from Eq. (3.16), the dependence of the ultimate strength on the alloying elements is elliptical, without their mutual influence. Taking into account the above, the method of gradient descent in the applied program for the solution of engineering-mathematical problems SciLab [8] obtained the optimum point max (σ_s): σ_B (8.27; 3.18; 2) = 1199.86 MPa.

Below are graphs of the dependence of the strength limit on the concentration of alloying elements (Figs. 3.12, 3.13 and 3.14). For each of the graphs, the value of the third factor was taken at the optimal level.

From the analysis of the equations obtained it is clear that the most strongly mechanical properties of γ-TiAl alloys depend on the content of niobium and molybdenum in the alloy. Noticeably affects the ratio between the amount of niobium and molybdenum. The representation of the results of the experiment by a polynomial of the second degree proved to be justified—a significant part of the nonlinear terms here is significantly different from zero. Since nonlinear regression coefficients have the same signs, the response surface is an ellipsoid, and its center is an extremum, with a maximum, since the regression coefficients are negative. As expected in the

Fig. 3.12 Function (3.16)
graph with value of Cr = 2

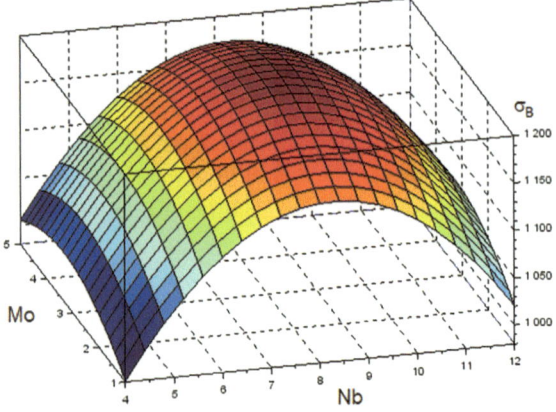

Fig. 3.13 Function (3.16)
graph with value of Mo =
3.18

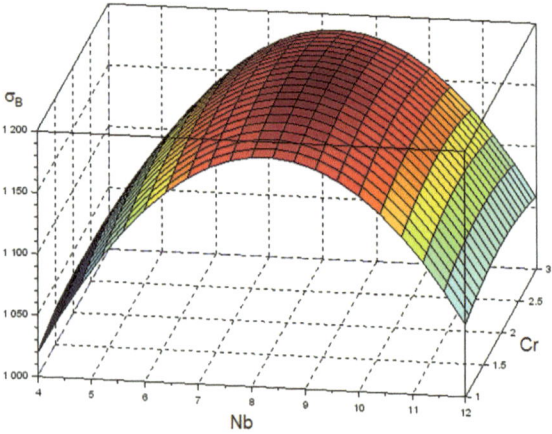

Fig. 3.14 Function (3.16)
graph with value of Nb =
8.27

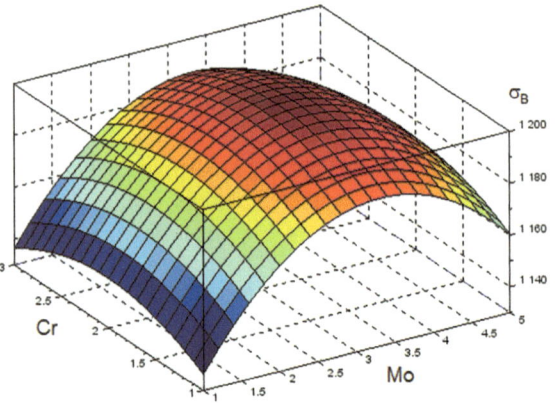

formulation of the problem, the optimal content of alloying elements of the analyzed composition lies in the experimental region, and near its center.

Equation (3.17) is an elliptic paraboloid. The extremum with respect to the third factor, taking into account the range of admissible values, is in the maximum permissible concentration ($Cr = 3$). Having carried out numerical optimization with respect to the two remaining factors, we obtain the following optimum point max (δ): (9.89; 4.26; 3) $= 4.63\%$.

Figure 3.15 shows the admissible values of Eq. (3.17), taking into account the optimization parameter ($> 4\%$). As can be seen from Fig. 3.15, the optimal value of the tensile strength lies in the rational region of the plasticity function.

Thus, the recommended optimum composition of the intermetallic γ-TiAl alloy, wt%: aluminum—30%, niobium—8.3%, molybdenum—3.18%, chromium—2%, titanium—the rest. (Formula Ti-44Al-4Nb-2Mo-1Cr).

Fig. 3.15 Function (3.17) phase plate with value of Cr = 3 (**a**), Mo = 4.26 (**b**), Nb = 9.89 (**c**)

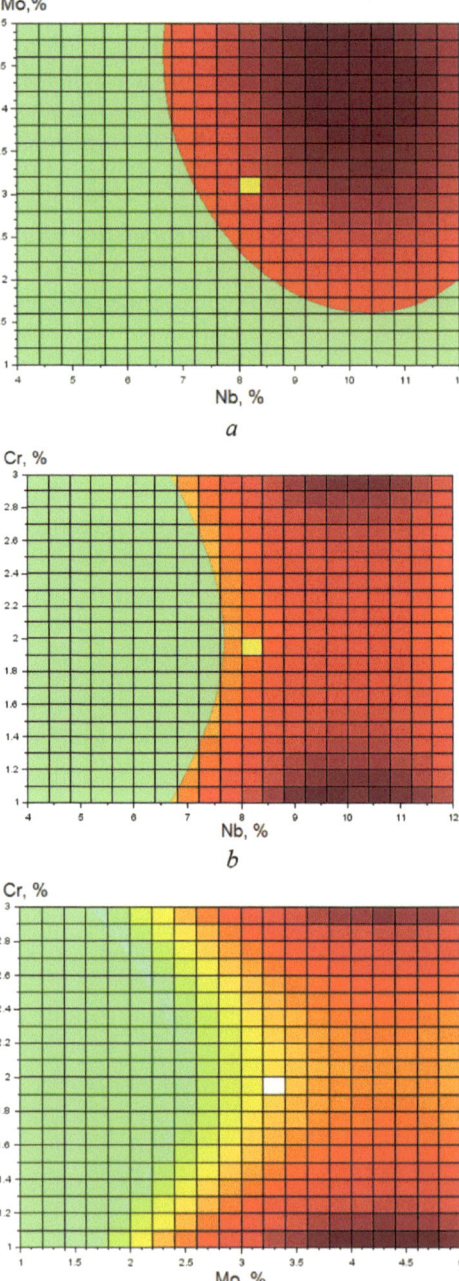

References

1. Imayev VM, Imayev RM, Gaisin RA, Nazarova TI, Shagiev MR, Mulyukov RR (2017) Heat-resistant intermetallic alloys and composites based on titanium: Microstructure, me-chanical properties and possible application. Mater Phys Mech 33:80–96
2. Leyens C, Peters M (2003) Titanium and titanium alloys. Wiley, Weinheim. https://doi.org/10.1002/3527602119
3. Lagos MA, Agote I, Gutiérrez M, Sargsyan A, Pambaguian L (2010) Synthesis of γ-TiAl by thermal explosion + compaction route: Effect of process parameters and post-combustion treatment on product microstructure. Int J Self Propagating High Temp Synth 19(1):23–27. https://doi.org/10.3103/s1061386210010048
4. Sereda B, Belokon Y, Sereda D, Kruglyak I (2019) Modeling of processes for the production of bassed alloys TiAl and NiAl in the conditions of SHS for aerospace applications. Mater Sci Technol (MS&T 2019), 2019, 137–142. https://doi.org/10.7449/2019/mst_2019_137_142
5. Sereda B, Kruglyak I, Zherebtsov A, Belokon' Y (2009) The processes research of structurization of titan aluminides received by SHS. Material Science and Technology, Pittsburg, USA, pp 2069–2073
6. Sereda B, Zherebtsov A, Belokon' Y (2010) The modeling and processes research of titan aluminides structurization received by SHS technology, TMS 2010, Seattle. Washington, USA, pp 99–108
7. Belokon Y, Zherebtsov A, Belokon K (2017) The investigation of nanostructure formation in intermetallic γ-TiAl alloys. In: 2017 IEEE international young scientists forum on applied physics and engineering (YSF-2017), pp 311–314. https://doi.org/10.1109/ysf.2017.8126640
8. Pavlenko DV, Belokon' YO, Tkach DV (2020) Resource-saving technology of manufacturing of semifinished products from intermetallic γ-TiAl alloys intended for aviation engineering. Mater Sci. 55(6):908–914. https://doi.org/10.1007/s11003-020-00386-1
9. Lapshin OV, Ovcharenko VE, Ramazanov IS (2005) Mathematical modeling of high-temperature of synthesis of nickel aluminides under pressure. Izvestiya Vysshikh Uchebnykh Zavedenij Tsvetnaya Metallurgiya 4:67–72
10. Lapshin OV, Ovcharenko VE (1996) A mathematical model of high-temperature synthesis of nickel aluminide Ni3Al by thermal shock of a powder mixture of pure elements. Combust Explos Shock Waves. 32(3):299–305. https://doi.org/10.1007/bf01998460
11. Lapshin OV, Ovcharenko VE (1996) A mathematical model of high-temperature synthesis of the intermetallic compound Ni_3Al during ignition. Combust Explos Shock Waves 32(2):158–164. Доступно на. https://doi.org/10.1007/bf02097085
12. Sereda B, Belokon Y, Sereda D, Kruglyak I (2019) Modeling of processes for the production of bassed alloys TiAl and NiAl in the conditions of SHS for aerospace applications. Mater Sci Technol 2019 (MS&T 2019) 137–142. https://doi.org/10.7449/2019/mst_2019_137_142
13. Sereda B, Kruglyak I, Zherebtsov A, Belokon' Y (2011) The influence of deformation process at titan aluminides retrieving by SHS-compaction technologies. Metallur Mining Ind (7):59–63
14. Belokon YuO (2018) Termokhimichne presuvannia intermetalidnykh splaviv. Zaporizhzhia: ZDIA; 2018. 220 s. (in Ukrainian)
15. Sereda BP, Kruhliak IV, Zherebtsov OA, Bielokon YuO (2009) Obrobka metaliv tyskom pry nestatsionarnykh temperaturnykh umovakh. Zaporizhzhia: ZDIA. 250 s (in Ukrainian)
16. Belokon K, Belokon Y (2017) The usage of heat explosion to synthesize intermetallic compounds and alloys. Process Properties Des Adv Ceram Composites II Ceram Trans 261:109–115. https://doi.org/10.1002/9781119423829.ch9
17. Sereda B, Zherebtsov A, Kruglyak I, Belokon' Y, Savela K, Sereda D (2010) The retrieving of heat-resistant alloys on intermetallic base for details of gas turbine Engine Hot Track in SHS Conditions. Mater Sci Technol (3):2097–2102 (MS&T, Houston)
18. Appel F, Paul JD, Oehring M (2011) Gamma titanium aluminide alloys: science and technology. Wiley, p 762 c

19. Kato M (2014) Hall-petch relationship and dislocation model for deformation of ultrafine-grained and nanocrystalline metals. Mater Trans 55(1):19–24. https://doi.org/10.2320/matertrans.ma201310
20. Shi Q, Qin B, Feng P, Ran H, Song B, Wang J, Ge Y (2015) Synthesis, microstructure and properties of Ti–Al porous intermetallic compounds prepared by a thermal explosion reaction. RSC Adv 5(57):46339–46347. https://doi.org/10.1039/c5ra04047g
21. Clemens H, Mayer S (2012) Design, processing, microstructure, properties, and applications of advanced intermetallic TiAl alloys. Adv Eng Mater 15(4):191–215. https://doi.org/10.1002/adem.201200231
22. Loretto MH, Wu Z, Chu MQ, Saage H, Hu D, Attallah MM (2012) Deformation of microstructurally refined cast Ti_46Al_8Nb and Ti_46Al_8Ta. Intermetallics 23:1–11. https://doi.org/10.1016/j.intermet.2011.12.012
23. Schwaighofer E, Clemens H, Mayer S, Lindemann J, Klose J, Smarsly W, Güther V (2014) Microstructural design and mechanical properties of a cast and heat-treated intermetallic multiphase γ-TiAl based alloy. Intermetallics 44:128–140. https://doi.org/10.1016/j.intermet.2013.09.010
24. Huber D, Werner R, Clemens H, Stockinger M (2015) Influence of process parameter variation during thermo-mechanical processing of an intermetallic β-stabilized γ-TiAl based alloy. Mater Charact 109:116–121. https://doi.org/10.1016/j.matchar.2015.09.021
25. Godor F, Werner R, Lindemann J, Clemens H, Mayer S (2015) Characterization of the high temperature deformation behavior of two intermetallic TiAl–Mo alloys. Mater Sci Eng 648:208–216. https://doi.org/10.1016/j.msea.2015.09.077
26. Wallgram W, Schmölzer T, Cha L, Das G, Güther V, Clemens H (2009) Technology and mechanical properties of advanced γ-TiAl based alloys. Int J Mater Res 100(8):1021–1030. https://doi.org/10.3139/146.110154
27. Belokon Y, Hrechanyi O, Vasilchenko T, Krugliak D та Bondarenko Y (2022) Development of new composite materials based on TiN–Ni cermets during thermochemical pressing. Results Eng 100724. https://doi.org/10.1016/j.rineng.2022.100724
28. Belokon Y, Hrechanyi O, Vasilchenko T, Krugliak D, Bondarenko Y (2023) Development of composite materials based on TiN-Mo cermets during thermochemical pressing. Int J Lightweight Mater Manuf https://doi.org/10.1016/j.ijlmm.2023.05.006
29. ISO 6892-1:2016 Metallic materials—Tensile testing—Part 1: method of test at room temperature
30. Kostolov OL (2003) Matematychne modeliuvannia metalurhiinykh system i protsesiv. Dnipropetrovsk: NMetAU. 124 s (in Ukrainian)
31. Nazarenko LA (2008) Planuvannia i obrobka rezultativ eksperymentu. KhNAMH, Kharkiv. 163 s (in Ukrainian)

Conclusions

The scientific concept of multicomponent alloying of dual-phase alloys has been developed, concerning the formation of a given set of properties, by improving the technology of obtaining finished products. The realization of the chosen stress–strain state made it possible to increase the uniformity of deformation, which contributed to the homogeneity of the structure and properties. An experimental-theoretical method for determining the relationship between the stress–strain state of a metal with a grain size d and the yield point σ_T is developed. This favorably affected the homogeneous structural state of the metal, the physical properties, the distribution of deformation, which in turn increased the uniformity of the structural state and properties of the alloy. The experimental dependences of the flow curves and the dependence of the deformation resistance on the temperature–velocity and deformation parameters of the hot deformation processes for integration into the DEFORM software complex are also obtained. Perform physical modeling of thermoplastic processing regimes on a Gleeble 3800 low-alloy, dual-phase steel with optimal chemical composition. Based on the results of experimental studies, analytical dependencies should be developed to calculate the deformation resistance at various temperature–velocity and deformation parameters of the hot deformation processes of the developed steels and to establish rational regimes for their thermoplastic treatment.

The developed experimental-theoretical method for determining the relationship between the stress–strain state of a metal and the grain size and the yield strength σ_T makes it possible to propose recommendations on the control of the processes of structure formation in the treatment of metals by pressure. This favorably affected the structural state of the metal, physical properties, the distribution of deformation, which in turn led to the optimization of the homogeneous structural state and properties of steel. Mathematical formulation and solution of the corresponding physical problem showed that there is an increase in the voltage of the metal flow in 1.13–2.0 times.

The developed technology of thermoplastic deformation of low-alloy steels of the type 10HFTBch ensures the production of a stable level of strength characteristics

S. Sheyko et al., *Thermoplastic Processing of Structural Metallic Materials*, SpringerBriefs in Materials, https://doi.org/10.1007/978-3-031-73896-8

(tensile strength (σ_s)—540 to 560 MPa, elongation (δ_5—25 to 29% in sheet rolling that meets the requirements of technical documentation.

The recommended optimum composition of the intermetallic γ-TiAl alloy, wt%: aluminum—30%, niobium—8.3%, molybdenum—3.18%, chromium—2%, titanium—the rest. (Formula Ti–44Al–4Nb–2Mo–1Cr).